# PLANTS: A MARVELLOUS GREEN CREATION

# PLANTS: A MARVELLOUS GREEN CREATION

Anthony Leopold van Ewyk

ATHENA PRESS
LONDON

PLANTS: A MARVELLOUS GREEN CREATION
Copyright © Anthony Leopold van Ewyk 2007

All Rights Reserved

No part of this book may be reproduced in any form
by photocopying or by any electronic or mechanical means,
including information storage and retrieval systems,
without permission in writing from both the copyright
owner and the publisher of this book.

ISBN 10-digit 1 84401 847 4
ISBN 13-digit 978 1 84401 847 5

First Published 2007 by
ATHENA PRESS
Queen's House, 2 Holly Road
Twickenham TW1 4EG
United Kingdom

Printed for Athena Press

*Introduction*

This book contains a gathering of data, mainly concerning the creation of plant life. So much goes on in the plant world that I felt it was worthy of compiling a book on the subject. The creation of plants turned out to be a world of its own, that of flora, and there exists another world within this one, namely that of flowers. The other world, which consists of fauna, is interwoven in the world of flora to quite an extent, as demonstrated by bees using flowers to make honey or cattle grazing on grass.

Upon study, it turned out to be most interesting as to how the botanical world developed. There is so much going on that it is definitely worth making a report of it all. After all, plants have been busy for millions upon millions of years on this relatively small planet.

All things within the green creation have the disadvantage of being rooted to the spot, which has urged them on to be that much more inventive in order to spread themselves worldwide. Consequently in came the other world, that of flowers, often emphasised as a world all of its own, with a flower being distinguished by a very slender long stalk, which, as the only hold on, connects these two living units. The other kingdom, which we belong to, is of course a very interesting one too, but plants are so important that we wouldn't be here without them. This is demonstrated, for example, in wheat, which is a crop of vital importance to us. Even if we only consumed fish or other sea creatures, their lives depend directly or indirectly on plankton, which is of green make.

Whatever the case, it all clearly illustrates the great importance of plant life and it is that kind of creation that made it worth writing about in this book.

## *Plants: A Marvellous Green Creation*

Plants are the link between Earth and the sun. In other words they involve the sun in their green life in a direct sense. The sun is therefore a VIP in its own way: a Very Important Place. Furthermore plants reproduce quite often with the help of insects fertilising their flowers.

Plants' profound depth in life is therefore, in many cases at least, simply to create the most beautiful flowers for others not within their kingdom, and many surely have been very successful that way. There again, animals entirely depend on plant life, and a cow grazing is just one example. Animals are fed by plants in one way or another. Flowers in that way show therefore that there is a link between them and us, and their remoteness is only relatively apparent. A flower is so different from anything else that it almost makes up a creation all by itself. A beautiful flower is a completely new 'breakthrough', while a human or animal is merely a shape that has changed due to evolution. Not only did plants develop flowers, often exceedingly beautiful, which they themselves cannot appreciate visually because they cannot see, they also usually provided them with a heavenly fragrance. Again, they cannot of course smell themselves – shape, colour and smell are not meant for their own kind, let alone more developed concepts like love – but they bring pleasure to others, not even belonging to their own kingdom, of whom they have no real awareness, since they cannot hear, see or talk. Yet despite these 'handicaps', plants sense those around them and it is this sensing which replaced familiar facilities. It is us who use the expression, 'say it with flowers', while for those to whom they really belong they are meaningless in that sense (as a bouquet or nosegay). Every part of a plant, from roots upwards, is functional, and this includes the flower. A flower blossoms at its best, yet it does not represent an ecstatic sexual display, and as far as plants themselves are concerned no flowery show at all attracts them.

There is an origin to all flora and fauna, and species are not only very different in their own world, but they are both very different as units. Some have said, 'There is no beginning and no end' – and there seems to be truth in this. Life surely seems to respond to that one. That sensing or homing in on something; that shivery feeling that one gets when an earthquake is imminent, has a lot to answer for and it plays a major role in life – whose home is a body. It makes deciduous trees lose their leaves in autumn long before that first cold of winter has reached it. Plants sense that animals are there, and, being aware, protect and prepare themselves.

Protect in the sense of developing defence systems consisting of poisons in the shape of stings, of which the stinging nettle is a clear-cut example. The word stinging nettle itself (*Brandnetel* in Dutch) warns us implicitly not to touch the plant. They prepare themselves, on the other hand, by means of blossoming, to be benefited by 'them' in their quest for more *lebensraum* (room to live). These defences come in quite helpful, in stopping larger animals from stroking against bushes or making a meal of some of the leaves. There are even trees which can switch on or off their 'poison tap' depending on how much they are being attracted by huge amounts of insects (caterpillars, usually) munching their leaves. And it doesn't stop there, because a tree also has the capability to 'warn' its neighbouring tree of an imminent attack, and the second tree can act accordingly. It is all a matter of sensing, which is beyond our knowledge. The elements of a plant are identical to those of animals in many respects. Only the appearance is fundamentally different. The plant feels the pressure of animals, especially insects and hummingbirds, and acts accordingly.

The issue of sensing is very important in both the animal and plant kingdom. Flowers especially came about by means of sensing, and there is often a very colourful creation. Additionally, there are in many cases male and female flowers which are entirely separated and the female ones have to be fertilised by those from the kingdom of fauna. Then there is that well-known fact of wind pollination, where the pollen from the male flowers is blown over to the female ones. Speaking of sensitivity, the same applies to a dog sniffing at the end of a tiny bare branch in wintertime along a footpath, or indeed sniffing at the end of a leaf of grass, finding out whether another dog has been there before. Furthermore, at the end of this sniffing business he marks the territory to make sure another dog will know he has been there. Again, it is a matter of super-sensitivity, just as in the case of plants. One outcome of this sensitivity in the extreme is a very colourful (flower) display and it is us, as I mentioned before, who use the expression, 'say it with flowers', while to the flowers themselves this means nothing. In other words, 'flower power' symbolises love to humans, but plants and flowers haven't got the faintest understanding of this concept. Plants cannot move, or only in a very limited space (like running-grass) so their moving on had to be established by those not from their kingdom, like seeds by means of birds' droppings or indeed transported simply by means of wind power. Earth needs the sun to create those 'green units', without which there wouldn't have been any species at all, and certainly no animal land life would have sustained itself without the provision of food in the form of plant life. The sun even goes one step further, as far as flowers are concerned, and a bee orchid flower for one is a very striking example indeed of the two worlds meeting.

A bee orchard has the shape, markings, colour, and even the smell of a female bumblebee. Thus it attracts a male bee, which after landing on it will fertilise it with pollen from another bee orchid, to result in cross-pollination.

Plants with green leaves, save for exceptions like cacti, are a link between the earth and the sun. For them the sun is much more important than just daylight and warmth. Their life processes depend on it entirely. Plants do not communicate like the animals, who have their different sounds, as in the case of birds singing or dogs barking. All the same, plants sense each other and one clear example is a dead-nettle looking exactly like a Stinging nettle, just to give the impression that it is not to be touched either.

Plants became globetrotters by means of keeping running, creeping and climbing. Bindweed moves even in a two-fold way, namely with stringy material underground and then winding up and above ground, troubling other plants. Strawberries do one better that way, and apart from seeds on the outside of their fruits, they grow plantlets on their runners.

Air transport turned into big business for plants, often in the most unlikely ways, like going it the burdock way. Furthermore there is air transport by means of tiny parachutes, which are so light in structure that they can float the passenger seed miles away; here again there is quite a variety on the theme. It goes even as far as possessing at least one wing, getting close to a true mini helicopter. Then there is air transport or flying within something (as humans do in an aeroplane), usually by bird. However, the 'disembarkation' isn't quite as glamorous. In fact it is pretty low-key, being essentially a dropping! A species in its life cycle experiences this sort of encasement twice over in such cases, namely as seeds, which is the pre-plant stage and contains all the data necessary for the plant to be within a body (bird or otherwise) transporting it to somewhere where it is deposited simply within the waste in the shape of a dropping!

A seed is like an egg, only it doesn't stay within the nest once 'laid' and its travelling results in very interesting reading. A seed is capable of enduring much more extremes than an egg in a nest and even makes it through the winter in many cases without any mother care, let alone brooding! A seed is a capsule packed with the data and food within, like an egg, but much more independent, and it turns into a sort of mini botanical spaceship quite often. The data within a seed is usually male and female, without necessarily causing interbreeding later, as a plant or tree. The ones depending on the assimilation process (plants) are aware of the other kingdom and act accordingly. They cannot move, being rooted to the spot, so they grow in height up to tree size, so high in some cases that even an elephant could shelter underneath.

One very interesting issue in the plant world is the rosebush. It is amazing that this plant, in shrub form so ugly and thorny, and all bare in wintertime, could possibly produce a creation as beautiful as a rose, often provided with a heavenly fragrance. It is just another contrast in plant life, which in its own way helps to make the outdoor world interesting for us, to say the least. To emphasise the ugliness of a rosebush, we prune it thoroughly just before new growth sets in early in the year, turning the bush into a woody stubble and furthermore opening up the bush or, rather, what's left of it, by more cutting back in the centre. This means that all inner growing branches are cut right back to allow more light into the centre of the bush. The shrub has now turned into barely more than a nasty stump, which could do quite a bit of harm to someone not aware of it in a mixed border. However there is a good deal more life left in the stumpy appearance than there appears. In fact, it is almost explosive. It will soon display a growth power bordering on the unbelievable. Short stems turn into long in next to no time, with flower heads to top them all soon enough. The latter are of course the famous and beautiful roses. The actual 'stump' has developed into a fully-grown shrub once again, all in the matter of a few months! Within such an outsize 'stump' therefore, there must be a roaring force of life – so to speak – to achieve the near-impossible in such a short space of time. Growth like that repeats itself a hundred times in a century, as long as there are gardeners to prune the bushes! I was one of these – a professional gardener – for quite some time, until professional artwork took over. It is them, the professionals, whose work has the effect of a seemingly raw 'crew cut' on roses during early spring. Were we to take a photograph of one such stubble after pruning, and then later on of the same shrub while flowering, it would be very hard to accept as genuine were it not for the fact that photographs don't lie. It all proves the point that there is somewhere a superpower which created – and still creates – all life, way beyond our knowledge.

Air plays a very important role in plant life as well. It has a three-fold meaning for plants: it conveys light from which they gather abstract green quite ingeniously; air as a moving mass (more commonly known as wind) serves airborne seeds quite conveniently; and last but not by any means least it comes in handy for breathing, plants depending on this function entirely as much as animals.

It is the more interesting since plants represent the stationary kingdom, rooted to the spot as they are, which tune into moving air. It is plants which built the first mini airships, fully equipped, demonstrated in the form of dandelion clocks (seed heads), just one very striking example of plants' ingenuity. It proves the point that we are all in the same boat as far as working things out is concerned, both plants and animals.

It was in the case of plants the more necessary to move from one extreme to the other, being anchored on the one hand and yet moving as pollen or seed consequently. It was wind power or flying insects and birds, which are a second aid in that respect. Air power is now big business in our society as well and we have come a long way from windmills with their wings attached to a structure still based firmly on terra firma. We don't even use the sky as a limit any longer and we venture into outer space.

Seeds manage to germinate more than once in the most unlikely places. We once had a small buddleia seedling growing, already a foot in height, until someone pulled it out from a small crack in the cement between two square stones on the footpath along the garage wall. It is most amazing that it ever managed to start life at all in such a dried-out crevice. There is another one now, again almost a foot in height, growing against the opposite wall, that of the house, along the same footpath, and the split in the concrete is barely visible. How does a large shrub seed, after landing at such an unlikely spot, manage to germinate there? It is completely baffling! Then there are all those plants growing on the vertical side of stone walls. This again concerns a functioning which arose from a sensitivity we know nothing about. We only notice the outcome, like the hypersensitivity present from seed to bee orchid. The sheer strength, the will to survive, displays itself also in the roots, lifting up tarmac in places. Then there is a lawn, where again growth has been put to the limit, but the boundaries are man-made. A lawn often concerns a green stretch where the boundaries are borders of flowering plants and shrubs, or neighbours' territory. Fields on the other hand are 'lawns' where the limitation on growth depends on cattle grazing and furthermore the farm's boundaries. The grass in such conditions (whether lawns or fields) grows so close that there is almost no room between their blades. It looks like a green sheet, but each grass plant is a green unit all on its own and there could be billions of them in one acre. Their root systems are relatively shallow, yet they brave droughts, except under the most severe conditions. Here again a form of green wildlife is where growth also shows itself in a superlative manner. The force in both cases, namely grass growth or that of a rosebush from barely more than outsize woody stubble each year, is just formidable. It is the kind of power that was necessary to make it all originate and create the wildlife which we see around us today. It borders on the unbelievable, yet we mow our lawns without giving it a thought. However, if we could put all the weekly cuttings of a grass-leaf together, it would turn out to be a huge amount – let alone over several years. An individual grass plant is just one of an innumerable quantity making up our lawn. We cannot imagine a rosebush as small as a grass plant, but we can make a parallel the other way round and compare 'cultivated grass' that is the size of a rosebush: wheat, which is also pruned yearly, in the form of harvesting. It too grows to full length in a matter of months, not including the ripening period. It does not start, though, as a stumpy plant, but from a seed, which is another very powerful life unit!

Insect-eating plants are also well known – but there is also a plant which 'eats' slugs – just as well, because slugs eating plants are alas all too familiar to us. It is the pitcher plant (*Sarracenia*). It emits a chemical to entice the slugs, who are attracted to the rim of the plant's leaf. A slug then slides within it, where there is a pool of sticky digestive juices. It struggles, trying to get out again, but that results in more juices being secreted. It is then 'absorbed' or in other words eaten by the plant. Again we see a sensitivity well beyond our knowledge. It would be worth producing them in their millions over the globe and selling them at garden centres.

Sensing starts before the sense of touch and it enables, for instance, a bee to visit a flower from miles away. A good question is: how long have bumble bees been around to give plants an idea of what to copy? A bee orchid is a rather new invention in comparison to the creation of the bumblebee! That a plant had the inner strength to express itself that way is completely amazing. It made life, that of plant and animal, as we know it today. As in the case of the bee orchid, it is not just a matter of evolution, but clear evidence of having to be able to gather data in order to materialise a structure through shape and colour. The principle has been present all throughout life on Earth, and such a creation as a bee orchid is proof of it in a most expressive manner. In that sense, evolution came second. In other words, a bee orchid is outstanding evidence that there is more to it. It made life on Earth what it is today. It has all been a matter of gathering data from day one onwards and evolution followed suit – that habit of gathering data was there from the start. Life comes from a planet where gathering data is common sense, but there remains the puzzle as to where this phenomenon originated.

The plant world is a green kingdom in principle, while the other living world is just a matter of variety on a theme. In both cases variety is enormous. Plants furthermore go underground as well as above ground; their lives depend on it. A tree is a green shape so much all of its own that it is hard to categorise it in the plant kingdom. It enriches the plant world to such an extent, its dimensions are sheer out of this world – some trees can dwarf an elephant. Trees come in a huge variety, but there is only one elephant image. Another outstanding example of data gathering in the plant world is of course a dandelion clock. Plants, apart from many having a controllable defence system in the form of poison, possess a special flower-opening schedule which is a matter of specified daytime hours. Bees know when flowers are ready to expect them – and that's a kind of precision work in its own way. It is that kind of ultra-sensitivity that resulted in life on Earth as is starkly expressed, for instance, between a bee and a flower.

It also resulted in the ability of a rare male butterfly to find a mate from miles away. It is a well-known fact that many flowers depend on strong sunlight to open fully, whereas honeysuckle, for one, is at its best during dusk. Bees among others know about such a refined timetable, while we have to put our nose to a flower to benefit fully from the heavenly smell. Only yards away from scented flowers we alas cannot smell the adorable fragrance any longer. For insects, however, incredibly long distances will do to pick up a specific smell. It is flowers' first method of attraction, the second being shape and colour and the third being the nectar within. That sort of precision work was necessary to achieve fertilisation in the case of many flowers.

Nature works in contrasts. There are annuals and perennials and, then again, a plant's life depends on underground and above ground activity. Many butterflies stay dormant in winter, in sheltered places, while plants whose flowers they visit have their own way of hibernating. Annuals will go to the extreme and survive as seeds in winter, whereas many perennials simply die off above ground.

Seeds get the harsh treatment. For them there is no soft bedding as in the case of birds, let alone parents giving them warmth. The only parental care they get is being provided with one sort or another of equipment, like being able to glide away with a feathery light parachute, just to make sure they will start life well away from home. On the whole it is for seeds a matter of 'it is all yours' from the start. There are many seeds which are to our advantage, as in the cases of wheat and, indirectly, apples, encased as they are, but here in both cases we have been helping things along a bit. Here again there are huge contrasts and to a grass seed a coconut is the size of a planet! Plants wouldn't have got half as far a field without the help of those from the animal kingdom. This too is clear evidence of a registering by plants.

Birds provide seed transport often, though indirectly. When life began on Earth, it had the inborn capacity to spread worldwide, which automatically put the question forward of whether it might have done so somewhere else. There just wasn't room for manoeuvre other than travelling through space by meteorite. For the most microscopic form of life to land here on Earth was the only option. That minuscule shape of life was however capable of gathering data, and that's part of it today as well. In the case of plants, they flourished – literally. It is this, a fertilised egg, microscopically small as it may be, which often produces a clever human being. It illustrates in one instance this world of life and its immense potential for expansion. It all started from a minute life unit and managed to develop in a huge spread of global proportions. It happened in two totally different worlds, as we know them today. Because plants cannot move, rooted to the spot as they are, they have to invent ways, like flowers, to ensure the moving world of animals will come to them to give them assistance. It went from strength to strength in both cases, being positively creative.

It is also true to say that it was plants who gave the 'green light' to go ahead with developing the other kingdom. It totally relies on them both directly and indirectly. Plants turned into both base and food for the other kingdom and many birds, for example, eat berries while they nest between branches. The green world is also a buffer zone, as it were, between water and land. Plant juices within are part of a complete food/drink intake, as in the case of cows grazing or caterpillars munching leaves. Furthermore, plants create plants often with the help of insects fertilising the flowers. It is therefore in many cases a case of no bees meaning no seeds, or, in other words, a matter of bee or not to be! Green life is also good at 'branching out' in its woody world.

Evolution came about due to hypersensitivity within, from the smaller creatures onwards. The green spread was to be part of it, up to tree size. If this planet had been bigger, life possibilities would have been much greater still. There is even wildlife on the poles. The variety of butterflies is enormous and it proves in its way the point of hypersensitivity, through the same quality bringing rare ones together. It is in itself a beautiful display concerning a degree of sensitivity way beyond our imagination. Life began here on Earth, however primitive in dimensions, with that kind of sensitivity included. It made life spread all over the world. It turned out to be also extremely colourful. We see it in birds, butterflies, fish, flowers and many others besides. It was that kind of strength within that landed here on Earth. Super-sensitivity was here from the start, in other words. It also shows from seed to beautiful flower, from seed to bee orchid or from fertilised egg to human being. Just to come back for a moment to one of the most ingenious creations in the plant world, namely the extreme precision work involved in creating a bee orchid, it is absolutely incredible and that by a... plant! Plants are somehow very keen on working things out. In other words, we are apparently not the only ones doing so! It is totally baffling. They are in fact so outstanding that they are shouldering us in that respect. Somehow there is that element present in all of us, plant and animal, that enables us to respond, and it is beautifully expressed in a bee orchid. It is that quality which made the living world what it is, a matter of 'message understood'. It would be interesting if we could work out the age of the bee orchid species along with its forerunner, the bumblebee. It is this sensitivity in the extreme again that made a dead-nettle look like a stinging nettle.

Plants developed from grass plant to tree size, with the former often growing underneath such a huge pillar, similar to ants and elephants in the animal kingdom, and that is in both cases not by far the smallest image!

It was all going to happen once life began on Earth, due to sensitivity in the extreme to data from the start. Somehow life's origin went through space at a perplexing speed as part of a meteor and again somehow crash-landed on this planet and survived, which in itself is a most extraordinary event. It is sheer wonder it all materialised. There was a starting point, but the total structure following, known as life on Earth, borders on the unbelievable. It all comes under the heading of natural – with its basis as the supernatural. It means that our existence is a mystery in itself. There is clear evidence that here again a supernatural power is present – no structure of ours. It is also that kind of quality which enables life to reproduce.

This hypersensitivity is everywhere in the animal and plant worlds and it is expressed three-dimensionally. Although fauna cannot live without the sunlight either, it is of a more indirect nature to them, namely as warmth and daylight, while for plants both Sun and Earth are of direct importance for their existence. Even for an earthworm, soil is only of indirect importance, unlike for the roots of plants. Even the richness of soil is important to plants. Life therefore is a trinity, namely the actual core and its dependence on two places. In other words Earth would have been left lifeless if the sun hadn't been such a 'close' neighbour.

Plants usually express themselves in green leaves. They are as a rule flat and paper thin, yet they are of vital importance for the other kingdom, whether directly or indirectly. Their green colour – chlorophyll – enables them to absorb sunlight, which is of elementary importance to the overall structure of a plant's image, whether it is tree-shaped or that of grass. It is this kingdom in which the assimilation process is of vital importance. It gained momentum from the start due to data gathering in both cases, that of plants and animals. In other words, life depends on two places to survive. It is that joint effort of those two heavenly bodies which was to be the environment for life to develop in a two-fold manner, the flora and the fauna worlds as we know them today. It has surely made this planet into a very special heavenly body. It was going to be a living world with its origin out of this world.

Its strength of spreading is enormous and there is a botanical example in the form of a specimen called *Titan arum* which grows no less than six inches per day. This shows the extreme strength which is involved. That goes for length, whereas a tree with branches all around shows it in width and a beautiful flower with its heavenly fragrance in depth. It is extraordinary, realising that a microscopic life unit was to occupy one day an area known as... worldwide! Gathering data really made the living world move and spread. It all started from next to nothing – not unlike the actual universe! In both cases it was explosive as far as the spread was concerned. Life in this sense turned out to be a silent explosion, so unlike that big bang which resulted in the creation of the universe. There was something big in the microscopic life unit millions upon millions of years ago. One section of it got attracted to the goodness of soil on Earth, and likewise to the light from the sun's rays. This eventually became the world of plants as we know it today. The other world is that of animals, which are also fantastic units, may they be plant-dependent, directly or indirectly.

Life gained momentum from the start in the animal and plant world, due to gathering data. Plants say it with colours in their flowers, which in itself is a result of super-sensitivity, and this kind of 'super' quality is part of life. It made plants and animals what they are today. The birth of life itself originates from a 'super' impact. The 'super' quality resulted in two living worlds, which are totally different in make. The evidence of this extreme superiority shows us the multitude of features in plants and animals.

It is also a 'super' quality that created the universe. It may well be the light, which has been part of creating life, since plants are still partly consumers of light, which is most extraordinary in its own way. Both the universe and life started from a minuscule unit and it turned out to be a silent explosion creating a living world here on Earth. It is also amazing the way the microscopic units developed into giant trees in some cases in the plant world – or elephants in the animal world. The difference between plants and animals is so huge that it in itself seems out of this world. Bearing in mind the fact that plants are depending on two heavenly bodies, namely Sun and Earth, they are extremely interesting. They are feeding on the light of the sun's rays and the goodness of the earth's soil. In other words, they penetrated in Earth's soil, secondly in the light of the sun's rays and thirdly in the other kingdom in many different ways, like flowers among others, of which the bee orchid is an exceptional case. Compared to plants, animals are the straightforward ones, although here too are exceptions, like a spider creating a cobweb to catch insects and a caterpillar changing into a butterfly eventually.

To put it another way once again: plants had it in them to incorporate Sun and Earth in their life cycle. This turned out to be a necessity to ensure life on Earth for them. Or to put it another way: life in plants' case is dependent on the sun and earth. It all boils down to the fact that plant life involves three places: the one which life came from; the sun; and the earth. All that was included in microscopic life units landing here on this planet. Plants prove that a heavenly power landed here on Earth and animals came second in this, although they too are a most extraordinary development. There is somewhere in the universe a hyper-power present and Earth has been affected by it. It is a mystery at least as huge as the actual big bang, the birth of the universe. Animals, however beautiful and complex they often are, seem to be straightforward in principle, so unlike plants which are incredibly complex in design. For plants it is a matter of partially living underground, while for us animals in general it means living on the ground. Earthworms and moles, among others, live underground too, but in tunnels, so are unlike plants. We are totally plant-dependent – we may eat a lot of beef, for example, but it still originates from a cow grazing on grass. The core of life that arrived here (or indeed, was created here!) consisted of two dimensions and one part got drawn to the sun's light and the soil of the earth for the sake of existence, while the other section used the green shape as a basis for its own living, directly or indirectly. In the latter case, a lion preying is a good example. Plants don't prey in their realm, but quite a few of them kill insects for the sake of consuming them. Plants, despite being rooted to the spot, spread worldwide all the same, only in their case they had to be so much more ingenious. They proved to be so in their extraordinarily colourful flowers and seeds – transported directly or indirectly.

Air means breathing for plants too, while the richness of soil means food to them, and their assimilation process, which is of vital importance to them, depends on sunlight.

It is amazing that modest paper-thin leaves resulted in plants the size of huge trees. It is the same quality that grows grass that covers vast fields. That microscopic life unit spread worldwide, expressed in two living worlds. Life is foreign in the sense that it doesn't originate from our galaxy.

If it weren't for the fact that plants are part of our living world, it would have been very hard to believe astronauts telling us they had seen a living realm elsewhere, partly consisting of the green rays of a far-off heavenly body. The living world of plants on such a familiar planet as Earth are nevertheless weird individuals. What made plants decide that light is of vital importance or, to put it in a more realistic form, what turned one part of the core of life towards the sun's rays?

Plants have leaves which are part of their basic shape. They attract the other world by means of their flowers among other things, as much as they are defending themselves from the other world with stings and thorns. It is clear that plants respond to the other world. It is a matter of being aware. For the other world, animals, it is easier.

Plants involve two worlds in their actual living – that of the sun and the earth – and something like that is bound to have happened elsewhere. Animals depend in fact on four worlds: the one where life has been created; the sun; the earth; and the world of plants. Plants are the go-between for animals. It is the green living world which is one huge mystery by partly living off a section of the sun's light. Plants are immensely important and we wouldn't be here without them! They are animal's food, while their own food is bizarre. Wildlife on the whole is one big joint effort. This kind of joint arrangement results also in fertilised eggs, producing offspring. It is wonderful in its own way. Plants put the accent on the fact that life is a mystery. If astronauts from another planet landed here, they would be perplexed about such a joint effort. They would find it especially hard to believe how plants survive.

Plants developed woody strength, which we use in building our houses. The microscopic life unit, which landed here on Earth, was to develop into two worlds, worlds apart from each other and yet creating one living world. Life would have been impossible on this planet if it had not involved these 'weirdos' known as plants. In short, we owe our own lives to plants. Something, somewhere triggered off an awareness in plants that they had to be there to enable us to exist. It may have taken millions of years before it was realised, but then life is closer to eternal existence than a planet like Earth and, were the end of this world in sight, then a life core would find its way to another 'Sun/Earth' environment. It is most unlikely that this combination of Sun and Earth is the only one in the universe, and therefore at least a plant-world could be around in many cases, with the green world absorbing another sun's rays in its light. It is all very far-fetched, namely the actual 'birth' of life, the journey to Earth and absorbing the sun's light by plants. The spread both ways from next to nothing is enormous, worldwide in fact. The sun's light was going to be of vital importance to us all, plants and animals, directly or indirectly; or to put it in other words: without it, there wouldn't have been life on Earth.

Plants roots are enigmatic too. They feed themselves on the richness of the soil down below and also anchor the plant or tree. Furthermore, plants are profoundly elementary in design, depending indirectly on the sun and the earth to exist. Being anchored to the spot also means they can't go for shelter and have to brave the elements fully or die. Plants link up with the 'outside world' where there is another heavenly body, the sun, emphasising the fact it is all universal. Something started somewhere and we are just one place where it all expanded. There is an intellect in it all and plants show this as much as the animals. It culminated in us. We are a creation within a creation. Paper-thin, small leaves resulted in growth to giant tree size. It is a silent explosion in its own way. Green leaves were also to be part of the basis of a growth of wildlife in general. It makes the sun extremely important in it all, whether directly or indirectly.

Insect larvae like caterpillars penetrate in plant life to such an extent that they create big holes in leaves. Slugs and snails are of course pests in that respect in gardens. It is definitely an unwanted penetration.

Hypersensitivity leads to flying as in insects and birds and indirectly in our case as a creation within a creation, that of Homo sapiens. Both Sun and Earth play a part in the very existence of life, and plants prove the point in a direct sense. In other words both bodies share their part in the actual structure of life. It is this triangle of Sun, Earth and the smallest form of life that made it all come true. Plants in other words need the penetration into both places. The environment of Earth and Sun proved to be ideal for the development of life here; possibly even in its creation something very small triggered off something very big, in the shape of two worlds. Life's birth happened on a planet like Earth or indeed on Earth, with all those potentials being part of it. Plants are strange by animals' standards, because they absorb sunlight; secondly they also absorb the goodness in soil; and last but not least they create flowers.

Earth alone was not enough to support life and the sun got involved in it in a very profound manner. It means that we owe life on Earth partially to the sun, expressed in a very complicated process. It is this which made life on Earth possible, a process animals are totally unaware of, namely the benefit of the sun's light to plants. It was to be of vital importance to both kingdoms, whether directly or indirectly. Where there is that combination of two places in the universe, like Earth and Sun, there could be life! That twosome is of vital importance to sustain two kingdoms, the one depending on the other entirely. It is amazing to say the least that plants, from such a simple beginning, created such an outstanding world. Animals achieved it by straightforward gathering of data without such elementary matters as the sunlight and the richness of soil. Plants use both worlds, Sun and Earth, for existence, and once that was achieved there came the second living world, that of animals, which use plants as a basis to exist on. In other words, the sun and the earth are of vital importance for plants to make a living and so in turn are plants for animals.

Without the existence of the botanical world, Earth would just have been another lifeless planet. If there were another Sun and Earth basis somewhere else and our own astronauts found it, they could plant grass there, but it would never form into trees even after millions of years, since green life came about by gathering data from the start.

Plants may not be too good at running away but they certainly make up for it in other ways, unknown in the animal kingdom, and they came up with flowers as their *pièce de résistance*. A flower is male or female – or indeed both. In most cases a 'home help' is incorporated in the actual vital process of fertilisation. Sex, as in the animal kingdom, is treated in a very private manner between two partners. Many plants however are in a sense much more broadminded that way and it is simply a matter of not just two but three 'partners'!

It is extraordinary to say the least that the life core found a basis like that of the Earth–Sun environment, which was to be ideal for development. It split up into two totally different life units. Was the life core possibly attracted to this environment of two bodies, both being very useful to green life? If not, it would have been a chance in a million in the huge universe to find this area in space! Attracting or getting attracted is a big issue in this living world. It must have been present from the start. This environment, that of Earth–Sun, asked for something special to happen. It is as if these two bodies were made on purpose to attract life's core. It was just right for plants, since in both cases it was a matter of material, either gaseous (the light) or solid (goodness in soil). Attraction might well have had to do with the actual structure of life's core and the same goes for ultra-sensitivity, both in an indirect sense. Both issues must have been there from the start.

Too much shade can kill a plant. It is as sensitive as that, while on the other hand it can stand up to the harshness of the climate, rooted to the spot as it is, while animals must go for shelter if it is too rough outside. The green would express existence in a triangle, namely: the sun; the earth; and the actual life core. It's of plant origin, and the animals followed suit, literally by means of leg power. That combination of Earth and Sun could be anywhere in the universe, sending out its attraction. Sensitivity and attraction have a great deal to do with the creation of a baby. The area here (Sun and Earth) was ready to accept a life core. All that was necessary was to create life and once that was achieved the three fitted together in an ideal manner. In other words it was ready here, as a very special place in the universe and the life-core did fit in perfectly. The animal world was going to be a world within a world – that of plants. It was just waiting to happen as far as the environment was concerned. Plants were to be a basis to enable animals to exist. There are exceptions: the bee orchid for one developed after the creation of the bumblebee. In particular, though, plants were to be of vital importance for animals.

We created television and computers among many other things, but plants created beautiful flowers. In both cases this was a display of ultra-sensitivity, or intellect. It is a bud swelling until it bursts open and expresses a display. Animals in it all are the most straightforward ones, although they too come in an enormous variety of shapes, colours and sizes. It is ultra-sensitivity, glorified in intellect, and we show this in many ways. It made us build houses and a great deal more besides.

However, we penetrate too much in the other kingdom, resulting in a decline for the green world and, furthermore, if our own activities continue to cause global warming and climate change, this interfering could have an enormous negative effect on the entire living world, unless we take drastic measures.

Cuttings are part of plants which, once out in the ground, can grow into totally new plants, shrubs or indeed trees. They express a way of life totally unheard of in the other kingdom. There is a flexibility here which is utterly perplexing. It is even possible to make a grafting that is a section of growth like that of a tree and to put it on to another tree, not at all necessarily the same kind, with the result that the tree plus graft could turn into that of partly pear tree and partly apple tree.

The actual creation of life itself and its spread on this globe is an absolute mystery. It is not unlike a virus spreading itself over a body, but only in a very positive sense. Both Earth and Sun are actually involved in the continuation of life. Plant life depends on a triangle, namely sunlight in general, the green image in its light, and thirdly the goodness within Earth's soil. Animals, from that point of view, just 'absorb' plants.

Life on Earth was originally life *in* Earth, and plants still express this to some degree. There were three necessities involved to make it all come true. Firstly the actual life core, secondly the sun's light and thirdly the richness of the soil on Earth. Minute species are still on this planet today and there is a huge difference between them and an elephant in the fauna world or, as in the case of plants, the growth of a tree. It makes it all more perplexing. The difference between plants and animals as far as their smells, colours and shapes are concerned is that with plants all three are starkly emphasised often, but not meant for their own kind, not even for any other member in their kingdom, while with animals these facilities are meant for their own kind, in the opposite sex.

There is a great deal of fine-tuning everywhere in nature. In fact this homing-in on, or sensing, manifests itself so often in wildlife, not least in the form of mimicry, that it must have played a major role in the actual creation of life. This extraordinary homing-in method is also very aptly illustrated in the life of a young swallow. Swallows in general, in their emigration over thousands of miles, prove the point once again that there is an ultra-sensitivity present which astonishes us. The young swallow flies back in the spring to exactly the same spot where it came from. It spends here only a few weeks before migrating to warmer regions. The latest brood hatches out during late August or very early September, when there is barely time to grow up fully before taking off on that incredibly long journey, full of hazards!

Flowers are created purely for the sake of offspring. Once this task is over they wither away. That takes place after the actual fertilisation has happened, which results in the production of seed. Seeds come in an enormous variety. It is amazing that an apple exists where there was once a flower – and only a short time ago at that.

Straightforward as it usually is in the animal kingdom to produce offspring, it is comparatively complicated in the plant world and often heavenly smelling into the bargain. All this was necessary to come to grips with reality – working on a theme of 'go and multiply', which isn't easy if 'go' isn't included!

Plants were found here once there was a life core, with the cooperation of both Sun and Earth in the actual life cycle. All the same from sea to land it was a more or less straightforward issue for animals in comparison to developing life from algae to flowering plants.

Sensing is also very apparent in animals and a dog is capable of finding its way back home over hundreds of miles in extreme cases. Apple pips owe their existence to apple-fall, like a crab apple, and again with animals in mind, which is a matter of sensing in a totally different manner. An animal once fully-grown is just another creature, while in the case of plants fully-grown means that more is to come in the shape of flowers, which are almost completely a living world all of their own, often emphasised by means of a long stalk. It is in the shape of legs that animals walk from one place to the other, while flying is another option for birds. All this is not available to plants, although they came quite close with their runners or air transport like dandelion seeds, besides many other ways. It is all a matter of sensitivity, just as in our case it went from cart to car.

There are no foreigners! That is to say we all are 'flora and fauna', made here from the start and therefore we are all proper earthlings. This place in the universe was ideal for that purpose and both places share their part in the creation. It is irrelevant in this whether life started in aquatic or terrestrial conditions, and it would have absorbed the goodness in water if its beginnings had been maritime. The spread of life is amazing if only we think of a tiny apple pip growing into a huge apple tree one day, with a large crop of apples hanging down all over the tree.

Plants in general respond in a three-fold manner to enable the continuation of their species, namely to the light of the sun, to the goodness in Earth's soil and thirdly to the other kinds of wildlife. It turned out to be a trinity in absorbing: namely the sun's light, the soil's goodness and thirdly the species of fauna, mainly flying insects. They need this three-fold input for their existence in as far the sun and the earth are concerned for all of them and thirdly the flying insects among others for their complex flowers. It comes down to the fact that part of the life core was improved in a very special environment, namely that of Sun and Earth – without it flora and fauna wouldn't have been created! In other words, no Sun and Earth meant no flora and no fauna. Animal life is entirely dependent on plant life, while the latter had the strength to go deep into matters, namely that of the sun and the earth. Only that combination made life possible in a two-fold way on this planet. Those were the potentials of the life core.

Sun and Earth were interwoven in it all from the start. They made a trinity and still do. Sunlight stands out as a speciality and the same goes for planet Earth as far as its soil is concerned. It is in both cases a matter of vital importance. Without it there wouldn't have been life on earth. It made plant life come first and foremost and secondly that of fauna, again absorbing matter of vital importance directly or indirectly, that of plant life. In both cases the kingdom is so huge that it is hard to believe an ant is still an animal, or that a tree belongs to the plant kingdom. Plants made the absolute impossible (to us) possible. They stay put and yet search for food successfully. They solved that apparently insoluble problem by letting their roots do the job of searching and manage also to gain sustenance in a way from the sun at the same time in their assimilation process of photosynthesis (a big word for a big issue). It is amazing, to put it mildly, that the most humble ones under us, the plants, came up with the most complicated process to survive. Had we been able to do the same, there never would have been starvation – even if the population increased a hundred-fold. As it is, we can't even suck up goodness out of the ground we stand on. Plants do this – sucking up goodness out of the earth, without causing pollution. Indeed, on the contrary, they bring back goodness into the ground at the end of their lives and not necessarily even then – leaf shedding sees to that – apart from the fact that they breathe out oxygen (so unlike a car!).

In the assimilation process, proof is to be found that being blind (as plants are fully) is no barrier to solving problems; the emphasis then comes simply on sensing. The variety of plant shapes is even bigger than that of animals and it went from grass to tree – and much more besides. On top of that – literally, in many cases – they created another world, namely that of flowers. Such a hugely variegated world as this has no parallel in any way in the animal kingdom. It is the more remarkable because plants are rooted to the spot and all the data has to be obtained from a given place. We ourselves travel all over the world, directly or indirectly, to get our information. We created the telephone, car, aeroplane and lots more besides. Plants breathe and throb just like any animal – *vive la difference!* – but they are nothing like an animal otherwise. This immense difference between them and those of the animal world makes us look on in awe.

If an astronaut could take along just one thing with him to another planet with intelligent life, it would surely be a plant with a beautifully coloured flower. The long flower stalk functions as a pipeline to the flower held on top of it. It also increases dramatically the flower show. It is all a matter of display in mind for a very good reason of vital importance. After all, flowers are there considering the next generation. Many are made very attractive solely for members of the other kingdom. It all started from less than scratch. Something triggered off something big in plants. It is a display that turned out to be a world all of its own. When a seed is transported by a bird and travels along with the wind, it only then germinates when this has taken place. It points the finger again at a joint effort, namely that of plant and animal, reaching a goal. To do so a plant must have sensed there are such creations as birds and made preparations accordingly. As it is, it is interesting to see what evolving can do and it resulted in our case, for example, in the computer from us starting as cavemen, among many other things, and no doubt this is only a halfway station and probably not even that, given time!

All of us on land not only live under the sun, but, what's more, live with the help of the sun. In that sense, not Earth but Sun is our life source in the first place. Plants have it both ways, namely feeding above ground and underground, all on the spot.

Life is one big mystery, although the sun and the earth have a great deal to do with it. Leaves are usually a paper-thin spread, absorbing the sun's light. They turn a tree 'stalk' into a solid mass known as a trunk, towering above any other growth or indeed life in the fauna kingdom. They are extremes of size, leaves and tree trunks. It is interesting to realise it is all a matter of extremes in spreading, like it is from apple pips to a huge tree, as it was likewise from life core to world spread. Being in one spot, as plants are, it is highly advisable to make the best of it, and therefore, with their roots, leaves and flowers, they show to be very active indeed in situ, unless dormant, which also happens throughout the animal world. The latter is a superimposed non-activity scheme. To be dormant could be in the case of plants either as seeds or as plants, whereby the plant supply dies off above ground, and, here again, like a stick stuck in the ground, striking is, in either way, unheard of in the animal world. The latter understand that under dormancy just sleeping may often be a sleep extraordinary. The sun is a life-giver in more ways than one and it is this extraordinary process which started to come in its own at a very early date in life on Earth (if not right from the start). It made life expand on a huge scale and it belongs to one of the many features of life – if indeed not the biggest display of them all. The dimensions of flowers are huge. Too big for us to imagine, but that issue of the plants obtaining food by involving the sun is bigger still. The issue becomes even more strange bearing in mind that it has been tackled by deaf, blind and dumb plants, not us, and it is an idea we would really have been proud of!

The life cycle of plants is the more interesting, because not only is the flower a very complicated unit, but a seed is much more an 'individual' or a personality than an egg, because it 'goes it alone', so unlike a bird's egg, which is badly in need of the mother's care. We take it all for granted, but, again, assume we are intelligent mentors here from another planet and it would then be experienced as great excitement, witnessing an insect visiting a plant by enticing it either on a specially made petal platform or a funnel-shaped structure, again specially developed to receive the guest, who actually completely disappears within the plant, that is, the flower. We would then realise that here we have two entirely alien groups of life (two worlds) who nevertheless come together or are still in contact with each other. A bee usually visits a flower as a complete stranger (and one cannot be more of a stranger than visiting from another world!) to assist in the fertilisation process of the plant, since mating is not possible in their case, immobile as they are.

Many plants developed in three ways: growth underground; above ground; and branching out.

A seed is not only so much more independent in the sense it germinates under its own steam, within reason of course, but it also results in plants moving along, very often being carried by birds. In short a seed is not just a capsule containing a great deal of data, enabling it to perform the near impossible, with contents powerful enough to create a large plant, indeed often a tree one day, but furthermore a seed 'moves home' though admittedly in a very indirect way.

The latter also goes to say that an egg, hatching, 'expands' for less than the volume involved in the development from seedling to plant, let alone tree! If there were roughly a parallel here, then a hatchling would turn into a super-giant bird, much bigger than a tree, from the starting point of an egg. Such is the silent explosion of a seed. The force present here is just formidable, to put it mildly. It results in a biomass seed multiplying its mass millions of times over in the case of trees, and that's pretty powerful stuff. That is only the length, while the depth in plants has its own formidable surprises, producing not just the flower as a *pièce de résistance*, but more – creating a flower that is an exact copy, in outline, of an insect (like the bee orchid), which is for that plant not just an ordinary stranger but in fact a visitor from another kingdom!

Sensing has a very great deal to do with the actual creation of life. It is expressed in a very grotesque way in the bee orchid, and a dog in some cases finds its way back home over hundred of miles. Their ability to sense, in other words, registers far beyond ours in that respect. It may well have to be with the life spark homing in on this very special environment, consisting of two most extraordinary bodies. This ultra-sensitivity also has its hallmark concerning many plants flowering almost two months earlier, due to climate change.

Not only did plants tune in and sense insects; it also went vice versa and there are astonishing cases, like a certain hawk moth with a tongue tube which has been made to fit inside a tropical flower tube of the same bizarre length. It results in a striking example of sensing both ways. Life and sense are one and there is definitely no nonsense anywhere. It is sense which brought life here on Earth, in the sense of homing in like a dog finding its way back home over hundreds of miles. It is plants sensing, which urged them to produce attraction, without which most plants simply wouldn't have materialised. It also gives strongly the impression that flying insects were especially created to please plants. Plants, somehow, defiantly realised that insects were there to be made use of and they improved their flower parts to attract them, while insects improved parts of their anatomy to obtain their beloved nectar. There are also others which assist in the fertilisation process, like humming birds.

Birds took flight but, in plants, life only then blossomed in its flower. There is nothing that equals flowers' unique and complex structures – not even any of our man-made creations! It is a built-in design and not made from borrowed material. It had to be really great to overcome immobility. Animals would be totally lost without mating leading to offspring. In fact, they would be a write-off. A flower's task therefore is an enormous one: not only should it embrace all the essentials leading to offspring, but it also ought to be included in the precision works to let others do the performing necessary to materialise fertilisation. To realise this, plants actually invite utter strangers inside their most intimate chambers – and that would be something to write home about, were we ourselves strangers here from outer space! The mind boggles at the thought of such a performance (of no romance and no mating) – yet just as natural and beautiful (almost more so, considering their flowers) as in our kingdom. There were no two ways about it, one would have thought – until plants knew better.

It is perplexing to say the least that plants even came to the idea to create flowers, let alone beautiful flowers, or flowers that look like insects! The latter especially shows clearly there is a directing everywhere – a steering towards a goal. Haphazardness would never have worked.

There were several options to put a title to this book, like, *Reading Plants: A Philosophical Look at Plants and Where the Plant and Animal Kingdoms Meet*. The latter part expresses the beauty of it all, since it concerns both kingdoms, and the bee orchid is accordingly an illustration par excellence. Plants are the green world, flora, or the stationary world, literally but by no means figuratively.

It is amazing to say the least that two worlds developed on this planet which have absolutely nothing in common. Plants have been just as creative and their variety is also immense. Creativity has been greater in the plant world from the moment beautiful flowers were created. The latter is usually on top of a long stalk to emphasise a world all of its own. The long stalk is just a green pipeline between the two worlds. The beautiful flower is the ultimate in inventions from non-borrowed material – and nothing in the animal world came to such height. Plants came up with more inventions like thorns and stings for the sake of defence – all because they can't move. It is also totally unheard of in the animal kingdom.

The way plants came into existence, namely with the assistance of Sun and Earth, was the only way to create a huge variety of life on Earth as two separate worlds. Plants are totally unique in the way they exist, completely unheard of in the animal kingdom. It is a matter of plants growing within the ground they stand on, on the one hand, by the means of roots, which are quite substantial in the case of trees. They are involved in the solid mass, given some exceptions, which is soil one way and the sun's abstract light the other way. So it goes from one extreme to the other for plants to exist all on the spot. It had to go to such extreme lengths to make life work on this planet. It was all part of it. The microscopically small start of it all resulted in two totally different unions, namely those of flora and fauna. It makes very interesting reading, but that's as far as it goes for us and the actual deep-down mystery is still there. Something extremely small triggered off something exceptionally big, an entire world twice over. It goes in length, width and depth, and the latter is hugely emphasised in us. We take it all in – almost. It is still far-fetched how paper-thin leaves can structure a massive tree trunk, to mention just one point. Data gathering was all built in from the start and in depth it came to creating beautiful flowers and one better still – us – though it may be that we are haywire sometimes quite a lot! It all started here, from almost nothing.

It is still amazing, to say the least, how the three came together, namely Earth, Sun and the life core, at one very specific spot in the universe. That alone is precision work in the extreme. It puts the emphasis on the fact that life is born here. The more so since it relies wholly on the presence of both earth and sun, as far as the making is concerned, for both places possess important things, namely the abstract in Sun's light and the goodness in soil on Earth. It is literally a chance in a million to find such abstract material as the sun's light or the solid mass as soil on earth somewhere else in the universe. Life is a trinity and that makes it a very slim chance to find it anywhere else in the universe. We too, animals at large, are totally plant-dependent, directly or indirectly. We wouldn't be here if it weren't for the plants – and they wouldn't be here if it weren't for both the sun and the earth. It makes plants very special indeed, and the deeper we go into their lifestyle the better. So if you do garden, have a deeper thought for plants.

Life after all these millions of years still depends on both earth and sun entirely, despite our living on just one. From leaf to fully-grown tree or from plant to flower are just a couple of examples showing that life's power is just enormous. There is a super force with unbelievable strength in all sorts of ways – even a rare male butterfly is capable of finding a mate miles away.

Had this planet been hugely bigger still, there could have been still aliens on Earth, intelligent beings we hadn't met yet. To us, aliens are plants. They may be in close touch with Earth but also with the sun. Both places were ready to accept a life core and, once it had been done, they both took part in the evolution, passively. For the rest, there is absolutely nothing to go by as to when, where or what started it all – and why! So if there is somewhere else – or indeed, in many more planets in the universe – a planet with weird plant life, there will be almost certainly also fauna, and possibly intelligent life.

Plants cannot talk (what a pity!) but they do express themselves in an enormous variety of ways. Plants go back to the origin of it all in their assimilation process, while animals in that sense are newcomers. In other words, plants today show us how it all started millions of years ago. It is this being rooted to the spot that makes them very stationary, yet they too managed to spread all over the world like animals! There was just that much more 'go' in plants, metaphorically speaking. They were more inventive. The core was a life unit which harboured an enormous potentiality, unheard of on this planet. It all became reality in a worldwide magnitude twice over. It has to be seen to be believed, even if the core originates from another body light years away from this one. The hugely complicated structure of that life core had to be made there, which only deepens the puzzle. It is amazing that plants are capable of making contact with another world, that of fauna, and the evidence of it is there in all sorts of ways. There is a mind at work in both worlds and it was part of the life core. It sounds more likely that it all started here with the extreme delicate structure – that of the life unit, which must have embraced such a mass of data from the start, microscopically small as it might have been. Fauna and flora are worlds apart in structure, basically. It is because 'something' dropped here on foreign ground, which was capable of creating (to us) very familiar shapes and structures that we know as plants and animals.

A flower is an attraction in shape, colour(s) and fragrance, often never to be noticed by their own maker – the plant! A flower is a show – and meant to be that way, emphasised often in its fragrance, yet it is also the most intimate part of the plant. A flower is indeed something very special and perhaps we ought to realise that much more. A flower is in fact many things in one. It is a 'food shop' for bees, where they obtain their nectar and pollen. It is also a magnet to them. It is a centre of reproduction, which is a huge issue. Plants grew sky-high in the form of trees by means of paper-thin leaf power or needle-shaped ones.

Then there is also the other extreme, as shown in ground-hugging plants. Once the idea of a leaf had turned into reality, there was no stopping of spread all over the world in a huge variety of ways. Life has in common with the universe that it was an explosive might, for starters. It is extraordinary that life here on this planet is just on the right spot, apart from the odd earthquake and hurricane. It is because plants had to come up with so many ideas to survive, being stationary as they are, that they are in that respect more interesting than animals.

It shows clearly there is a mind at work and the displays are only too familiar. Flowers are an expression par excellence in that sense. Beautiful flowers especially seem out of this world, but then so is the image of a plant! It is in flowers that plants went for striking colours with the intention to attract members from the other kingdom – for a very good reason. It is the more amazing, since the green colour of plants is only there for the functioning of the assimilation process of photosynthesis, which results in absorbing sunlight – purely for the sake of living. How could such an extremely delicate life core have stood up to the most extreme elements of outer space? It would have been another mystery besides. The crash-landing wouldn't have done much good either! That life unit must have been something very delicate, which eventually developed into two entirely different worlds. Colourful flowers are just a 'message understood'. The registering has been there all along in life on Earth. It is because of that being included that it happened.

Evolution is development by involvement and this is where sensing comes in. The force in nature is an incredible force in the extreme. It makes an apple pip develop into a huge tree one day, full of apples in season. We ourselves started life from so extremely little to go by all those millions of years ago.

Flowers are three-dimensional, namely: colour, shape and inside, while butterfly wings are a display of beauty. Plants were there already during the period of prehistoric animals. Although some plants live hugged together, like grass in fields, or forests, they have nevertheless a very isolated way of looking after themselves. Plants' elementary shape is the leaf. It is a shape that enables them to exist. There is nothing like it in the animal world. The question is what triggered off the assimilation process (and one just can't imagine a bigger question). It is worth two living worlds on Earth, the animal world depending, directly or indirectly, on the green world. They are by far the oldest in evolution, well before prehistoric times and are still here in their billions, although we destroy nowadays a good deal of the green world for the sake of 'development'. It would be interesting in the extreme if we could hear what plants had to say about these things! It is this leaf image that created plants worldwide. They made room for the animal world who, without plants, wouldn't have stood a chance. It is this mysterious assimilation process that triggered off so much on this planet, namely a living world twice over.

Plants not only made room for another entirely different world, but they also created the world of colourful flowers. The lifestyle in the green world is different in the extreme, compared to animals. A flower is an image on top of an image (one plant) or a creation twice over, all for a very good purpose. Plants are in very close touch with the sun and the earth, and this intimacy with the other kingdom. Plants' origins therefore are mainly inorganic, in the sense that they are created by sunlight, while it is a matter of being mainly organic in animals' cases. It results in two entirely different worlds, yet plants attract members from the other world. The big question is why life turned into two big worlds and not just one – that of plants, who look after themselves! As it happens, it turned out that once plants were there, animals had to follow. It is as if it had been worked out that way. It was all programmed from the start.

Many of our inventions are most ingenious, like an aeroplane or a computer, and we are a developed species. We are nevertheless unlike a colourful flower, which is in principle a new creation.

Plants came up with the idea of paper-thin leaves, while with animals it is just the opposite with their bodies as a rule. The life core started to develop into two basically totally different creations, soon after its own development. If the life unit were created on another planet, then there must be life there as well. What is more, it would have repeated itself time and time again. In that case we are not unique at all. It is plants that are the straightforward ones, in this sense. They use the sun and the earth to make a living. It is animals that are the puzzling ones. While plants build from strength to strength by means of their leaves, animals developed a head as the main object. They have a 'head start'.

Once the leaf was invented (there may be exceptions to the rule like succulents and cacti), the variety on the theme and the growth in general was colossal. It is especially shown in places like fields and forests; it shows to be a magnitude beyond any calculation. Alas, a decline, especially in the animal world, is due to us! Plants spread in greater quantities (like fields) worldwide than animals. Although plants spread all over the world, save for uninhabitable places, there was still plenty of room for animals and they live in quite a few cases just on top of the green mass, as demonstrated by cows grazing. It is amazing to say the least that such a highly complex creation as the assimilation process was there from about the beginning of life on Earth.

Wildlife soon got involved with the sun's rays. It meant two living worlds were to be created consequently. In both cases the quality and the quantity turned out to be enormous. There may be no heart beating in plants, and no brain, but they worked things out for the better all the same. It is extraordinary that a life core was a structure which had the potential to create two totally different living units of world size! It worked out to be a 'just so' process and there never was a wildlife catastrophe. It all fitted in perfectly well, until we came on the scene.

Our way of life would have an extreme impact on the overall structure and will continue to do so unless we take drastic measures. It is the more remarkable, because we are the brainiest (or so we think) of the lot! There may be a good deal of killing taking place mainly in the fauna world, but that is all for a very good reason, just to survive. We may not sympathise with it, but there is no other way. Apart from that, both the flora and the fauna worlds are perfection in their own kind, which all started from a microscopic life unit. What could have started off a development, which turned out to be so huge it became worldwide without making mistakes? Even we, the most powerful of them all, have our errors occasionally! It is most probably just a coincidence that such a life core and its environment, that of Sun and Earth, met. But the fit is too perfect for that. Somehow those two universal places were just made to get involved with a life unit.

The big question still remains: why did it happen? It is the more remarkable because it happened on just the right spot and not in the Arctic or on a desert.

What is a plant? A plant is a green living thing that exists by an assimilation process and is usually based on paper-thin leaves. An animal, on the other hand, is as a rule provided with a head to work things out. A life unit found here a rough world of stones, mountains and sand or clay only, save for the watery environments. But then, if it started off in an ocean, how did it manage to become terrestrial as well in a two-fold way? Plants have a greatness we cannot altogether fathom! It is a creation beyond belief. What triggered them off to make use of the sun's rays for existence? It is just incredible! Such was the supernatural strength of the living unit when it sorted itself out. More than just planet Earth was going to be part of it all, which eventually lead to two living worlds, completely different in their way of living. Not only was planet Earth a barren mass when the life unit was created or arrived here, but the climate must have been very rough as well. Without this most remarkable feature of the assimilation process, there would never have been life on Earth. Indeed, what triggered it all off? Life has been around for millions of years and it doesn't show any sign of 'old age' at all, although our interference cause interruptions!

Plants have a big 'say' in it all, silent though they are. They are truly aliens to us, since their lifestyle is very weird. They have no vision, yet it is them that got in touch with light for the sake of survival! It was also them, being blind, that 'observed' bees among others and satisfied their requirements. It was them, with no eye for beauty, that created the most beautiful things on Earth, namely colourful flowers! Having a close look again at the origin of wildlife, it is amazing that it ever happened. It is sheer marvellousness. In the two-fold split that followed, plants are even more miraculous in their existence than animals, since they actually involve part of the sun's light rays in their daily lives. They are indeed a profound mystery twice over. It is the assimilation process, mysterious as it is, which is vital to plants, but also indirectly to animals – including us.

Animals have their bizarre ways of expressing as well. It is not just the bee orchid or the way plants trap insects. There is an insect in Pakistan whose head seems to be at its tail end and vice versa. So precisely is the seemingly reversed issue expressed, that any bird would be fooled by it – and to make doubly sure, the insect starts its flight backwards!

Plants made good use of wind power as well in their spreading of seeds. They created structures especially adapted to float seeds well away from the mother plant, and a fully equipped dandelion seed is an outstanding example in that respect. It stands out in making use of wind power that plants too possess a super quality. It is amazing that plants too, despite being restricted in expressing themselves since they have no heads and therefore no faces nor eyes, managed nevertheless to create a variety of items to their advantage, glorified in colourful flowers. Thus sense reigns in both worlds. Seed data that develops into tree concerns a magnitude of unsurpassed greatness, leaving us way behind – and not just physically.

They both, the universe and life on Earth, have in common that they are expressions of huge dimensions. Life started off here on Earth as a minute unit and now it has reached a biomass of billions upon billions of tons, including us.

Plants exist in two ways. One is abstract, using sunlight, and the other is a matter of taking up the goodness from the soil. It is totally unheard of in the other kingdom. This involvement is planets apart and it makes plants extraordinary to say the least.

It went that far, beyond anything to create life on Earth. A bee orchid is a miraculous structure and is as marvellous as asking a blind man to dress, in make and colour, the same as his sighted friend.

As much as animals roam, plants stay put. This alone puts the two kingdoms worlds apart. When the rooted to the spot basis was introduced, all sorts of fundamental issues had to be invented, which resulted among others in a floral display. In flowers, plants actually involved life from the other world. They, bees and many others, became much more than just 'shareholders'. For us it is just a bee visiting a flower, that's all – but what a performance is really going on!

Going to such an extreme as creating immobile units, like

plants, was the only way to make room for another kingdom on land. Planet Earth alone was apparently not enough to feed plants successfully, and it is because of the involvement of the other heavenly body, the sun, that plants show themselves to be capable of mastering the impossible, like creating colourful flowers. We cannot fathom the issue completely! A creation that involves partly the sun, though indirectly as in the case of plants, is bound to be extremely powerful, as shown in creations like sky-high trees, and above all in most ingenious, colourful flowers. In many cases there is merely a pipeline, which a stalk represents, as a go-between from the actual plant to the flower. We are plant dependent, directly or indirectly, which results in the belief that we are not wholly earthlings either. It makes plants (including trees) in that respect also an out of this world creation. A good question: did life have animals in mind when it developed plants? It certainly couldn't have come along the other way round.

Life is growth and once it comes to full maturity, it is flowering time in most plants' case, which is so much more than 'ordinary' sexual development, as it is in the case of animals. It turned out to be a super-extension of 'ordinary' plant life. An extension which led to a new world almost completely on its own. There is just a stalk left between the two worlds, those of plant and flower.

Plant flower and seed go their own way and as delicate as a flower happens to be, a seed is relatively tough, as a rule.

We take it all for granted, but plants growing to the size of trees for one is so amazing that in fact words fail to express it profoundly. A lone, massive, fully-grown tree is a matter of green growth, usually of gigantic dimensions. That's as far as trees are concerned, but a beautiful big flower on a long stalk is an invention in depth and a display one wouldn't have expected humble plants, headless as they are, to be able to create. It is all because once plants realised – somehow – that there were species in the animal kingdom which could assist them, like flying insects especially, to extend their world, they created plants with the most beautiful flowers purely for the purpose of attracting them to fertilise their new creations. Plants went to such lengths (and depths) and it worked most successfully! A flower is such a heavenly and colourful structure with its fragrance. Wisely it is as if it was all well planned before the actual arrival of life on this planet. It shows that plant's green environmental effect on Earth was necessary before another living world, that of fauna, could develop here. At the same time plants would make use of species like flying insects, belonging to the other world.

A rose bush is either thorn or flower, with the bush acting as a mediator. A first-class extreme on one plant!

A flower doesn't look plant-like at all and is purely for species from the other kingdom, which complete the fertilisation process in some way; many do this the wind-pollination way.

Attraction plays such a big role in life that it must have been present in the actual 'blueprint'.

The two worlds spread worldwide, but they never became overcrowded despite the numerous species making up both worlds.

It is a fact that planet Earth was just right for plants and they feed themselves on the goodness in soil, and plants were just right for animals in the case of nutrition. Plants spread worldwide as well, which in their case is the more so an outstanding feature, because animals have legs or a pair of wings to move along. It looks very much like plants were pre-planned for the sake of fauna. It all happened just so from the start and no catastrophes occurred. Overpopulation is to our disadvantage and we cause 'disruptions' in the worlds of fauna and flora.

It is all the more remarkable, because animals rely on plants and to a certain extent so do plants on animals. Both worlds respond to each other. Such life must have been there from the start. Even a crab apple with its edible volume round its core or a dandelion seed with its clock are clear evidence there is reasoning everywhere. It gives the impression, quite clearly so, that it was all pre-planned, because it all fitted in precisely and the variety on the theme in both worlds is enormous. It is just humans who have wars. The deeper one looks into it, the more the mystery deepens, as far as both worlds of flora and fauna are concerned. It's amazing that plants without any mental faculty whatsoever managed nevertheless to 'invent' ways of self-defence, like thorns and stings, among many others, but, above all, flowers. The latter is simply for survival's sake, as complex as that process may be.

Calamities never happened between species or indeed between the two living worlds until we arrived and there is, alas, that slaughter known as war. Plants grow in many cases shoulder to shoulder, so to speak, in the cases of fields or lawns or indeed forests, but they express a 'happy family' effect. The same goes for animals, which often move in large flocks or grazing herds. The enormous variety in plants and animals settled on earth quite peacefully, taken overall in a 'just so' manner. It gives the impression of a huge *fait accompli*. There is no animal that comes close to a tree with apple in mind, which makes plant life more remarkable. It gives a topsy-turvy impression, making it more difficult for us to puzzle it all out. Instead of doing so, we get ourselves more stuck in life's cobweb.

Plants, looking at them from this angle, are life's brainchild. It is their extraordinary extensions that put them into the front row – and stalks with a flower head on top of it just emphasises the point.

What made young life on this planet absorb part of the sun's light to survive? Plants started life in a very far-fetched way by involving the sun directly in their daily lives – just to make room for a second living world, that of animals. The big question is: what living force was urged to get involved in something so much out of this world, which relies on the sun's rays for life? It concerns a hugely complex issue, that of a photosynthetic involvement. It puts plants in a very special position.

Life arrived here on this planet (wherever it came from) with one very specific element standing out, namely to absorb, by photosynthesis, some of the sun's light, which would sound simply too far-fetched if it wasn't for the fact it actually happened! It is something like: if it can't come out of the length, it has to be out of the depth.

Nonetheless, it is amazing that plants have been able to hold their own, because it has been so close to asking the impossible – namely to stay put and yet to stay in business. According to us, the two don't go together. One simply has to make one's moves, especially nowadays. That involves a huge amount of travel, usually. In a way a plant and a factory (a plant of works) are not all that far apart from each other. In fact they have a great deal in common, but then we are very sapient, while plants have to make do with sense only. Yet they attract and distribute all from one and the same spot. That's business in a big way! A plant itself cannot travel, but a seed as independent and, small as it is, has that potential. A beautiful flower is worth a great deal more than even a bird of paradise. Immobility has a lot to answer for in plants' case but they overcame the difficulty marvellously and their flowers are a source of delight to us and many insects and others, quite apart from the fact that blossoming has a very deep-down meaning. It is amazing that two totally different living worlds were to be realised from one very small life core.

They have absolutely nothing in common except sense. Life is a super-power of extreme qualities. This is expressed in animals as much as it is in plants and we ourselves are an extraordinary display in that respect. In short, plants had to happen. However, it was asking much, if not the impossible, to make room for the other kingdom of fauna, which is provided with heads. It is the more remarkable they followed suit, because they are literally worlds apart in creation. The two may have nothing in common basically, but flora attracts fauna to quite an extent, and they even catch very small prey in some very special cases, in a most ingenious way. It was as if plants were made especially for what was to come – fauna! Attraction is a big abstract in the plant world too, but in their case it is for a third party. Attraction in plants is even more important because it has to outweigh immobility.

A fruit is a most ingenious creation and meant solely to deal with the huge problem of immobility. That is the sole purpose of an apple, and that's why it is delicious for many of us in the animal world. It was originally the crab apple, and the pips were dropped with the excretion after consumption well away from the original tree. There are two apple cores thrown in our garden on a small clearing between trees, and they are gone the next day every time. So there is still a demand for them.

Plants spread in height, width and length. In height with the trees, ever reaching towards the sky in many cases, in width with their branches extending and in depth with trees' roots underneath the surface which could be again quite substantial. There is even a fourth dimension, which is in depth, figuratively speaking, with their flowers.

Plants penetrated very deep into matter, like soil and light. Moreover, they also went the furthest, creating their flowers, resulting in a way of life almost too far-fetched. There were two buddleias growing in a most peculiar position. They were still seedlings and only about one foot in height. Their leaves were already of normal size, though. The extraordinary thing was that a seed landed in both cases at a spot where it was sheer impossible to germinate. They were about a few feet from each other. It nonetheless happened in both cases, almost opposite each other at the side of a very small paved footpath with cement between the stones. This narrow pavement was between the garage and the house and was just over three feet wide. It all fitted perfectly and there was not the slightest crevice between either wall and the path. One of the single-stemmed shrubs grew all the same on the garage side, just above a very narrow cemented gutter and the other one of about the same length was against the house wall and here again there was hardly any split at all in the cement against the wall. However, both buddleias looked very healthy and would have made much more of growth if they hadn't been pulled out. The root system of one of them was very shallow, dense, a couple of inches wide and about six inches in length.

The perplexity of it all is the sheer unlikeliness of a seed germinating in such circumstances, and furthermore surviving in a most unhealthy environment. There is here clear evidence again of the supernatural force behind it all. It was the way plant life found a hold on this planet. It concerns a super-power that borders on the unbelievable – even so tiny as a buddleia seed may be! Plants are therefore clear evidence of an incredible might present involving both the sun and the earth in their daily living, with paper-thin leaves as a basis. It is that super-power which was embraced in the life core. It turned out to be a silent explosion involving all the world.

Fruits too show themselves to be an extension of the actual plant. They are also a three-dimensional creation, like flowers, only in a different way. Fruits and flowers materialised with fauna in mind.

Plants prove time and again to sense as much as the animals. A clear example is again expressed in the holly bush with its leaves being quite prickly at the edges on the lower branches, but being smooth higher up where it is out of reach of herds. Obvious to us, but here we are in the plant kingdom!

Life found its super-challenge in a flower. It is a display of supernatural design, which is a great deal more than just development. A flower is where the puzzle of life turned tangible. Therefore there is no completely satisfactory definition of a flower, or for life for that matter. A flower is where plant meets animal, often in the shape of an insect. It is the place where plant meets life's demand, which turned out to be huge in their case – so unlike boy meets girl. It is the place where immobile plant meets kinetic bee. A flower is to bypass immobility, and to be a fertilisation station. It makes a flower blossom with greater emphasis than a normal sex-life expression in the animal world, however beautiful their plumage. A flower is there to attract fauna and not its own kind, which makes it in a sense a weird object, but wonderful also at the same time. It gives the impression that a flower, strictly speaking, does not belong to a plant. It is, as it were, too 'way out' for that. Plumage of one sort or another can be lovely, but it is still definitely part of the animal and it doesn't come with a stalk for one. Flowers, consequently, make plants rather special in the living world. It had to go to such lengths (and depths) to make it all work.

Grass's habitat may be like a field grazed by herds or even lawn size, but nevertheless there is a type that does grow up to tree size and is known as bamboo, which is especially well-known in the tropics. It shows again plants' extreme might. It is all a matter of super-power that originated from the life core period. It is a pity in a way that this planet, known as Earth, is only very small, otherwise the spread of life in general would have been so much bigger and possibly even to such an extent that there could well have been still aliens to us here on this world!

A plant, just because of its immobility – which is such an unlikely feature, life-wise – had to develop all sorts of extraordinary – to us, delightful – ways to make ends meet, so to speak. Reaching out is expressed as branching out, which plants are extremely good at, and secondly there is spread without direct movement involved as in the case of many seeds. Plants are green for a very good reason, however far-fetched. It just couldn't have been more elementary than involving two places in the daily life cycle. Without that dual incorporation, there simply wouldn't have been plants or indeed animals. In short: no involvement of two places in the life cycle and there wouldn't have been life on Earth. This utter complexity was the basis of it all. Animals from that point of view are more straightforward and followed a path of evolution, however fantastic the outcome all the same. Life on Earth was created twice over, namely as a kingdom of plants and consequently another kingdom, that of animals. It is those kinds of super-qualities which were centralised within the actual life core. It was indeed a very special union, namely that of the life unit and planet Earth. It gives the impression there was nothing coincidental about it. There was too much precision. This therefore gives strongly the impression that a super-power is responsible for it all.

Overcrowding is no problem in the plant world, as expressed in fields and lawns, and, moreover, agriculture thrives on it. Life is a super-force and it is alas us, the brainiest of them all, who cause a contrasting effect on the structure. It has a topsy-turvy influence. Plants are the 'go and get it' stations (stationary as they are!) and we all obey. The big question is how plants were to know bees have a sweet tooth (or tongue, literally).

The reaching-out effect in plants developed to such an extent that it could be divided into four distinct groups: reaching out for the sun, absorbing a specific part of its light; roots; fruits; and, last but not least, flowers. All that while being almost as immobile as a stone; it is a fact even more strongly expressed in plants by being rooted to the spot.

A plant draws aliens to its premises. It even convened its premises in its flowers for this purpose, simply sensing other life. It is a display of remarkable sensitivity – which puts even that of the sensitive plant in the shade. The latter, however, is most remarkable and it folds the leaves together which are touched. This ultra-sensitivity, which is clearly expressed in the shape of a colourful flower, resulted originally in the living image being attracted to the sun's rays and making partial use of them, extremely far-fetched as that might have been. It led to the creation of plants, which at the same time got interested in the goodness in Earth's soil. Such a colourful flower is almost completely separated from the actual plant and a long slender stalk is the only evidence of a belonging. It was created with insects in mind, going for the goodness within. The huge potential ties were there in super-seed fashion, once there was a life unit on Earth. It was going to be a worldwide spread in a two-fold way. A flower on a long stalk like a dandelion is a light. It is proof of sensitivity in the extreme. It is evidence of another world, the third one in life on Earth. A flower, in other words, has a much deeper meaning behind it than to be picked with others to be put in a vase for the sake of sheer beauty, according to us, the (apparently) most highly developed creatures on this planet. A flower is sheer beauty expressed in a tangible way due to sensitivity in the extreme, but the reason for its creation goes much deeper than beauty and it is there to attract assistants from the other world. Minute as it may have been, that life unit was at the same time also a super greatness of the first order. We came about due to evolution, while a flower is a creation all by itself, like plants themselves. In short, a flower is a creation within a creation, like a plant is a creation within a creation – that of the life unit.

Plants are exceptional in the extreme, by involving two places in their way of life, while animals, although also fantastic creations, came about by evolution.

Plants created flowers solely with pollination in mind. 'Three's a crowd' is not only acceptable, but a necessity in a plant's case. As it turned out to be highly complex, plants had to be there to sustain animal life.

We came along all in good time. It is as if it had all been sorted out beforehand! There was no mistake made over millions of years. Life has in common with the universe the fact that both are super-powers. As it was not enough that life was actually settling here on this planet, next came the sheer unbelievable might of plants which involved no less than two bodies, Sun and Earth, in their daily life, and in both cases for the sake of nutrition. It would have been sufficient for only plant life to cover the planet; animals are just extra, the more so since the two kingdoms have absolutely nothing in common basically. Why did it all happen, when a simple life form would have been sufficient? We top it all and make it possible to review it. It all took its time, but it was well worth waiting for to see us appearing on the scene. We can oversee it all except the very beginning. It is just too colossal to imagine and we can't take it in.

In both cases, the universe and life, the items involved go in their billions. Plants exploded in growth to such an extent that it is very hard to take in that a sky-high tree still belongs fully to the plant world. It went to such lengths that trees are almost a world of their own. It has also been near impossible to accept that our apple pip becomes a fully-grown tree one day! It is such an explosive matter, which started life on Earth, not unlike a seed growing into a big plant before long. This environment was apparently a perfect base for starters, and the sun had its say in it all. It turned out a great success, as we know only too well today, millions of years later. There is no sign of 'coming of age' whatsoever after all this time. To the contrary, there is an everlasting effect apparent. In short, life is a perennial.

One day, more monkeys may turn into intelligent beings. It all started from next to nothing, but then so does a very small seed and there is an enormous amount of data within.

Once life settled here on Earth, it spread like wildfire in quality and quantity. Nothing was left to chance and it all fitted in just so. It shows the mighty potentialities of wildlife. There is evidence of a *plan de campagne* everywhere due to following up sense. It gave reason to a following suit from the very beginning. In short: it had to happen. That puts the emphasis on an omnipotent force being present. Whatever the case, it very clearly gives the impression it was all pre-planned. Even in a bee orchid, super-might is exposed in plant fashion. Super-sensitivity has been there from the very beginning and it plays a big role in it all. It turned creation into reality. It was like a 'big thought' put in black and white, writing-wise. It was like building a house: first and foremost there has to be a floor. As it turned out to be highly complex, plants had to be there in the first place to sustain animal life. It is awe-inspiring to say the least, that Sun and Earth had to be involved, however indirectly, to create us ultimately! In other words, the two bodies play their part in it, plant-dependent as we are. Life soon developed in a highly complex manner – in plant life and its end product – again a highly complex design, capable of creating numerous features.

Life as a life core came quite possibly with the big bang and settled in places like the sun and earth environment. That it involved two places shows the fact that it was highly important. It also started from next to nothing, like the universe. Now there are billions upon billions of species and the grass plants alone in a lawn are very numerous.

Sensitivity in our case is expressed in all sorts of variations and most of them very ingenious – so unlike a bee orchid for one, its sharp fragrance and sweetness which is more than humans' mental strength.

Once more: what is a plant? A plant is a highly complex creation due to involving two places in its daily life cycle. There are billions of plants and they are of vital importance to animals. Plants are 'mission impossible' made possible. It is as if it was realised that the plant, however complex, had to come first before there was a possibility of creating another totally different kingdom, that of fauna. A plant is mainly tissue, while an animal is usually a body in a true sense. In both cases though, there is a great deal going on inside. It is only more obvious inside a body and makes it more mysterious within a plant. Certainly a leaf couldn't be flatter, usually showing no body, but it is definitely not a nobody!

Plants too survived by adapting, which is so much more obvious in the case of animals where there is a 'central station', usually in the form of a head in command of everything. Somehow, plants too were under control all the time and it turned out to be an enormous manifestation. Clearly there is a great deal more to it than plants suggest, being made up of juices and fibres. Somehow within a plant there must be a main body and neither leaves nor stems nor branches are anything to go by, while a flower is usually a temporary affair (in some cases flowers pop up for only one day in the life of a plant). That on the other hand a plant manages to produce a flower at all is in itself clear evidence that there is much more to it all than just juices and fibres.

Whatever it was, plants were to be the 'chosen ones', to make life possible at all on Earth for animals.

In a plant, attraction blossomed. To be attractive is so much part of animals' life, but a flower for one proves the point that sense is also within them. For plants it is common practice in many cases to draw alien life, mainly that of insects, towards them to complete their life cycle. It is amazing – more, it is perplexing – that plants became aware of animals, insects on the whole, as shown in visiting their flowers. It shows quite clearly that plants too master sensitivity. Animals come in a huge variety, due to evolution; plants too show a great variety of species, but they not only developed sky-high in trees, they involve both the sun and earth in their daily life. Roots are such a might all by themselves that they became pot-bound often in the cultivated types. We may take it all for granted, but it is well worth giving it all very serious thought occasionally, because it is especially plant life that thrives daily in a most extraordinary way.

Animals, like insects, made plants create colourful flowers. It somehow registered. Their beauty to us is in fact only a functioning in flower fashion, a matter of attracting members from the other kingdom. It is the way that ultimately leads to plants' offspring.

Plants were to be the most complex structure ever made on Earth and that includes all our inventions. Where we have our limitations, plants have more mileage. Plants went much deeper into matter than animals, and it shows! A flower was to be the outcome of this combined effort, namely the underground pipeline network of roots transporting vital juices, but primarily this absorbing potentiality of the sun's rays, enabling them to survive.

A plant therefore goes by helping itself in ways unknown to the animal world. It goes literally to the bottom of things in obtaining juices from the soil, furthermore absorbing light in the assimilation process and, last but not least, obtaining help from insects or others in the fertilisation of their flowers.

Plants went out from one given point, and it is truly amazing where it all leads to. To live is to move – not so in the case of plants! It is beyond comprehension for us to live on the one spot all throughout life, from cradle to grave. Plants have a completely different approach to living than animals, and what is natural in our case, like raising a family, doesn't even exist in plants. Attracting is in many plants even more strongly expressed than in the animal world, all because plants can't move, so it really has to be an existing display in colour, shape and fragrance in the form of flowers. Thus they reach members in the other kingdom. So much effort had to be put in to overcome this most difficult feature of immobility, and in came colourful flowers as a complete new creation within a creation and, furthermore, seeds travelling in an indirect manner.

Plants turn absorbed light from the sun into tangible matter, via photosynthesis. All land life depends on this weird feature of the assimilation process. The source of it all is the other heavenly body – the sun. Plants started off from nothing, with only sunlight to feed on, and it is still their 'main course'. It made them an abstract absorbent force. It usually results in paper-thin leaves.

Plants are just an extension from the start itself; an image from next to nothing. It resulted in an absolutely huge spread worldwide. Fauna too had their way of spreading all over the earth. We see the evidence of both cases all around us, but we cannot absorb the start of it all entirely. Abstract sense has a great deal to do with it all, at least as far as plants are concerned and we are all plant-dependent. We put abstract thought to tangible structures, like mobile phones and aeroplanes. We call it brain power, which is again an abstract quality. A plant in general went one step further and created the flower.

There is plenty of evidence that plants respond just like animals. It registers in both worlds. Plants in other words are not just growth only. There is much more to it and it shows in outstanding features. Grey matter, brainwork, is an outstanding feature in humans, and possibly green matter is not all that remote from grey matter. It certainly made a huge variety of plants and trees all over the world. In other words, the abstract plays a big role in life on earth. Plants show themselves to have been also quite imaginative and adaptive, and therefore a great deal of initiative went into it.

Plants have their 'living room' in a literal way and definitely in an outside manner, save for some indoor plants. The green world has to face up to problems on the spot. There is no getting away from it in their case. Plants prove that life's environment stretches out beyond that of Earth, by including the power of the sun.

Very mundane issues which are so straightforward in animals as a rule, like eating and mating, are quite a complexity for plants. Plants could have gone it alone, although forms would have been very different in comparison to the general familiar outlines we know at the present time. There would have been mainly mosses and ferns and, of course, wind-pollinators. It is incredible, the utter harshness that delicate seeds endure in the winter. Stresses put upon them must be of a huge magnitude, tiny as they are. This is just perplexing!

It is hard to swallow that trees are in fact gigantic plants. It shows clearly what might was involved to create the plant world. Then there are of course big flowers, which prove the point only too well over again. In their case it is of course not in length but in depth. 'To be or not to be' – that is the question, but that turns out to be quite a complex one in plants' case and, besides sunlight and soil as cooperators in their very existence, they need wind for the ordinary fertilisation process with the plant itself as growth in the centre of it all.

Even a flower is quaint in a way, especially at the point where it involves fauna, all masterminded by a plant. It is almost as far-fetched as aliens assisting us by just existing.

A flower is more remote from a plant than a caterpillar from a butterfly. The former presents a world all of its own, save for a stalk, which is the only indicator it is plant-orientated. Plants and animals are so very different; it is amazing the two ever coincided on our small planet. It is however not only home for both, but they even intermingle, despite their difficulties. Sense plays an enormously important role in plant and animal life alike: its origin is alas totally unknown. Even the creation of the universe made sense. Catastrophes on a huge scale never happened, like planets colliding.

We owe our lives to plants. We originate from the animal kingdom and we are all plant dependent. Therefore without them we would be lost. Plants wouldn't have materialised if the sun hadn't been there at a very specific place in the universe. It was all very precise, the way this all was planned out. Plants were the breakthrough to establish animal life.

What is a flower? A flower is an expression of sense materialised; the very issue that played such an important part in realising life on Earth. Its origin, that of sense, is a big mystery. The beauty of a flower, however strongly pronounced, is merely subjective to its real quality. It is a matter of abstract which is behind it all, but then so are the sun's rays, which are plants' daily feeding source. It's clear that the abstract has a great deal do with creating life. After all, even evolution is basically abstract. It is indeed the sensing that shaped life on Earth in the first place, and a beautiful flower merely put the emphasis on it.

This writing wouldn't have materialised without having profound thoughts about it all in the first place, and that's what made life materialise, first and foremost. Life is therefore like a gigantic beautiful flower realised, save for some mishaps. Sense therefore was the forerunner of it all. It is in other words sense that plays a major role in it all. Once dead, no matter what, it disappears to nothing all in good time – which is abstract again! It is also sensitivity that has a great deal to do with making the huge variety of animals that we see today. In other words sense plays a major role in it all from the very beginning. In short, life made (created) sense.

A flower is like a very colourful advertisement. They were advertising their wares long before we came on the scene. A flower is a fleeting image of the supernatural. A flower is a creation within a creation. All this huge variety in shapes, sexes and colours in the plant and animal worlds is due to hypersensitivity, which also developed our brain power.

Plants reach bees, figuratively speaking, which are the only fertilisation option in many cases. Bees are the missing link in this, which gives the first impression of being highly improbable. Wind pollination is already an extraordinary state of affairs in this, but bees! For this, a whole range of causes and effects had to be arranged, like colour, shape and smell, and above all a reward in store, after a search has been successful – and all this finds its focal point in what is commonly known as a flower. This semi-dying-off process in plants, which turns trees to skeletons and results in perennials becoming almost invisible before the really cold weather sets in, is a much more drastic way of acting, as preparation, than simply hibernating as animals do. It is as if nature had run out of ideas as to what to do in such impossible circumstances. It is just a matter of staying alive, when there is no such thing as being on the move or going for shelter, in the case of plants.

Sense got in touch with this environment of Sun and Earth, and creation followed suit, like a flower, which started from nothing. It is creation within creation. Plants were a major might in it all and they are abstract-based. They turned the abstract into tangible matter via a highly complex assimilation process, and this was to be the basis for all life. It certainly proves the point that the abstract plays a major role. Plants and flowers are abstract-based. Plants do get the message when too many caterpillars munch their leaves and they act accordingly, as in the case of some trees that increase the amount of acid.

A dandelion may be a commoner and is usually considered a nasty weed with a pen root. It is, however, a very interesting plant, since it grows in rosette fashion and quite cleverly makes a very long stalk with a flower on top to be pollinated by the wind, and later on it spreads its seeds with little parachutes over a wide area. It is all a matter of responding. Sense is behind it all and that goes for creating the universe as well as life on Earth. It is in neither case a haphazard issue and consequently no calamities in either case have ever take place – or it must have been a very local occasion, like the extinction of the dinosaurs. It was a matter of spread on a huge scale in both cases. There are billions upon billions of individuals involved on Earth if one only considers the amount of plants involved in a lawn. The big question is, what mastermind is behind it all?

It is also beyond our knowledge what power is responsible for turning abstract matter into a tangible feature, like paper-thin leaves created by the incredible assimilation process. It is this assimilation process which feeds animals indirectly. There is a link between the two mights, that of the universe and life on Earth, and it is all due to sense. A plant comes about the abstract way, as did a flower originally, due to 'mind over matter'. Flowers are there for a very good reason and the beauty of many is 'just by accident'. Life has been from the beginning an affair of mind over matter and a flower is just a very clear example of it. Plants materialised only to be followed by a second, more important world, which created us as the *optima forma*. Plants somehow got in touch with sunlight in a profound way. Life on Earth materialised in a far-fetched way, as plants show only too clearly. So much had to happen before it reached us. Alas it is not all clear-cut in our case! Plants required something so little (some sun rays), yet so far-fetched, to create life.

In other words, plants are a most unlikely development, but they were necessary to create life on Earth in all its greatness.

The other way plants make a living is also most extraordinary, namely by means of absorbing the goodness out of the ground they stand on. Their root system is so powerful it turns them in many cases pot-bound when planted in a pot, and roots can lift up flagstones or even tarmac as they get bigger. Such is the force life shows in many ways. It is a weird way of living as plants show, and is responsible for a huge two-fold way of life on Earth. The living world is therefore an indirectly Sun-related issue. It is very interesting to realise there is such might in sunlight that it created a living world of plants! It is also interesting to realise that if there hadn't been such a special quality in sun rays, life would never have happened on Earth, even aquatic life, being dependent on green mass like plankton. A thought is there only for a moment, until it is put down in writing, and then it is there for ever. Something like that created life on Earth. It is like an architect building a house: a thought turned tangible. Plants take it one step further in their assimilation process.

In fact, they went one step further still and created from nothing, millions of years ago, the flower, for a very good reason. They too had it their way. Plants came here to stay, without roaming around like the animals. They just stand there, it is true – but in a far from idle way. It is interesting to note that it is plants, not animals, where so much had to be worked out to make everything function properly. There is no display more attractive than that of beautiful flowers and we, the most expressive species, put them in vases, these creations materialised by blind plants. There was originally 'no body', so to speak – and again it is plants which still prove the point clearly, by depending on the abstract. Plants are still nobodies in the true sense. Plants prove in a very basic manner that the abstract can turn tangible. This creation is a result of the most peculiar process of assimilation. Somehow it worked, and in a most successful way! That highly complex process was to satisfy a whole living world, life on Earth. The process is far-fetched, but had to go to such lengths to make life viable. The impossible, to us, was realised. It was a greatness of invention that all creation depended on. It makes creation of the living world superlative.

Plants originate from abstract matter twice over, namely via the assimilation process and evolution. Plants start off from seeds, but what's the beginning of a seed? To plants, sunlight was to be much more than just light. Needless to say, after all that, abstract matter has a very great deal to do with life on Earth from its origin onwards. It is in fact almost impossible to avoid mentioning abstract matter as the basis of it all. The most important features, like the assimilation process, evolution and the original creation of a flower, prove the point that the abstract is responsible for the existence of such things.

Making sense is behind it all from the very beginning, and a bee orchid emphasises the point by even reaching the other world that way! It shows in no uncertain terms how abstract matter is involved in the creation of life on Earth. Survival is for the fittest and this is poignantly illustrated in the thorns of a rose bush. Nettles' stings may be of a different make, but have the same aim – beware – which has basically a great deal to do with being aware. It is the latter which is shown in a particularly colourful way in a flower, which is abstract matter turned tangible.

Attraction is another form of abstract matter, as part of life on Earth, and it leads to so much tangible creation. It just shows how important abstract matter is to life. We all depend on three of these dimensions: the photosynthesis process; evolution; and attraction. The assimilation process is the major factor in life on Earth. The abstract turned tangible happened in life at least several times, as shown in the assimilation process, the original creations of flowers and evolution. It is a case of sense over matter. Indeed, it is the flower which started from point zero, millions upon millions of years ago.

The saying goes: an apple doesn't fall far from the tree. However, a tree's offspring is not usually found underneath it. That is because such a tree banked on the fact that many an apple pip would enter the stomach of an animal and be deposited somewhere far from the original tree, within a dropping. A flower makes a plant what it is really worth. Had such a flowering plant been found only recently by explorers, it would have made news of a world-shattering impact. Now we take it all for granted. That impact would have been even greater, had it been realised that such a colourful object possesses a strong element of attraction, which allures life from another world to assist it in the struggle for survival.

Plants are better equipped to look after themselves, stationary as they are. They are capable of drawing assistance towards them all the way from the other world (mainly in the form of insects) which is necessary in the most important stage in their existence, which leads to seed formation. Furthermore, plants are a contradiction in terms since they are on the one hand very earthly creatures, being rooted underground, yet they feed themselves also from in and out of this world on the other hand, via the assimilation process. Plants, by being green, have a most uncanny way with the sun. Their lives depend on it completely. It went from abstract to tangible matter all the way except for the fact there must have been a core of life present at the very beginning, which found here excellent accommodation.

A plant started from abstract matter originally, as did a flower millions of years ago, and animals developed via abstract evolution and therefore all life must have been very close at its starting point from abstract matter, and it turns abstract again after death, eventually. The sheer tenacity of life is most amazing. It is that sort of strong willingness that made life find a foothold on Earth at the very beginning. It concerns a might far beyond our knowledge. The smallest point will satisfy as an environment and be sufficient to accommodate a life core in the shape of a tiny seed.

Despite the fact that plants are very restricted in their ways of life, they developed a richness nevertheless, equalling at least that of the animals. They spend all their lives on the spot, and all are basically green save for some exceptions. In their case, the famous mountain of progress had to be drawn towards them, which is clearly the reverse of what occurs in the animal world. Also, a head is usually an obvious extension of the body in the case of fauna, whereas a plant and its flower have barely anything in common, least of all the colour. Plants had to be extra inventive to survive and develop in the world. For them it was a matter of being self-contained on the one spot provided. They surely went to the limit with their flowers, which moreover are often quite intricate structures themselves. Flowers are plants' invention, in fact sense is behind it all to a great extent and that goes for plants as well.

Plants started from something like the tiniest possible germ, and the major part was abstract involvement. Animals started life on Earth likewise. The plant/animal feature is a creation of a joint effort, which opened the sluice gates to an endless range of species all still firmly under the heading of life. Plants went even deeper into matter, by rooting themselves into soil. That bond couldn't have been a closer one between Earth and life, and it went in their case even from life on Earth to life in Earth. To make do with the one spot given in life, from cradle to grave – and even flourishing that way – set plants apart. They highlighted green as the go-between, leading ultimately to the sunflower. It furthermore shows life in plants as not purely earthly. That so much was to be had from 'outside' (outer space) in itself indicates life to be a force that is extraordinary.

A skylark feigning a broken wing and walking away from its nest in that manner, a moth that looks like a bird's dropping on a leaf, or a dog covering up its own dirt are just a few examples out of many, clearly showing it is all a matter of sensing. Life means contact made. It is its trademark. This has been so all the way – and even most probably too during pre-life, when it was about to happen. It is attraction that matters and without it life falls apart. The 'boy meets girl' version has a parallel of sorts in plants, and insects may not bring the two together, but they play a similar role. Paradoxically, because they can't move, the attraction element is much more strongly developed in plants than in walking or even flying creatures. Life itself is, in a way, a 'foreign body'. Wildlife simply grew and gained momentum that way from a given point.

The start of it all couldn't have been more modest. Wildlife today in its huge magnitude as biomass is the outcome of a silent explosion. A moth might have to travel literally miles to meet its mate, which is something in the order of a boy being attracted by the perfume of a girl a village away! That's strong stuff but also very true. In order to go to such lengths, obviously much has to lie behind the image of animals. It even turns almost into the ridiculous in the case of that moth, in our eyes. This is a matter of attracting beyond limits known to us. It is also life's very own. This 'beyond limits' motif is a universal issue, leaving limited Earth way behind. There are many signs in that respect. It far outreaches its own boundaries, expressed in attraction and mimicking, among many other issues. For a child, a creature in mimic is something to laugh at. For an adult, it is simply perplexing. Mind over matter made us masters over wildlife. Mind over matter also made plants flourish, indeed made plants in the first place. Plants are almost impossible creations, since their way of life is simply too far-fetched to be taken seriously; yet still we see the evidence all around us of their existence.

All the different shapes in animal and plant life and the amount is enormous. They came about by sense, which is abstract matter. The same quality, which makes it all, happened with the life core billions of years ago. It resulted in this core falling apart into totally different sections. They have absolutely nothing in common apart from being life. Grass proves again how firmly rooted it is, since cows graze a field without completely pulling out so much as one plant in the process. Life is basically a 'from outside' effect, materialising time and again and clearly demonstrated in the bee meets flowers example – and that 'outside' part could do with a second hard look. There must be a good reason, for it has been incorporated in living to an almost ridiculous extent and, although a plant doesn't look anything like a bee, the two are extremely close in intimate matters and that 'relationship' goes even further in the case of the bee orchid.

Features in nature are manufactured by having had a 'feel for it' in the first place. It has been a matter of increased brain power which enabled us to get airborne, literally and figuratively. It is a similar sensing which managed to put a flower on top of a plant and, therefore, purely elementarily, it is all a matter of making sense. Although plants and animals differ enormously, they are equals when it comes to 'feeling one's way'. The next illustration is quite a striking example in its own way.

The comma butterfly not only looks like a crumpled-up old tree leaf when at rest, and displays a very tattered edge when its wings are spread, but moreover the stark, whitish comma shape on its otherwise brownish wings looks unusually like a crack in such an old tree leaf. It gives it so much the more realistic camouflage. It is again evidence of sense, which shows in both plants and animals.

The difference between an animal and a plant is of an absolute quality. It is a kind of intense greatness in itself – or, in other words, an out of this world quality. It means that all nature is in fact supernatural. There is something in the make of life which is as enigmatic as that of the origin of the universe.

Once again: what is a plant? A plant is a most weird form of life that needs part of sunlight to exist, that builds a root system which absorbs goodness out of the soil. It also attracts small flying creatures, like insects, all the way from the world of fauna via a display and a fertilisation that comes in the shape of a flower, in many cases to assist in that all-important process. A flower is clear evidence of making sense by plants.

Animals, although a huge gathering worldwide of all sorts, came about by evolution and lack these peculiarities necessary for plants to make a living. Animals wouldn't have been realised without those 'weirdos' known as plants. It turns life into one huge complexity and the sun got involved in it all. Plants had to have that bit of 'extra go' in them, so to speak, since it all had to be achieved from the one spot, while animals could simply move from A to B. Plants, in short, are green evidence of so much. They went all the way from algae, via mosses and ferns, up to trees, both literally and figuratively – and that is a very great height. In plants life went in depth via roots and the performance repeated itself in us, metaphorically. Plants' variety show is one with a big difference, since there are numerous participants giving a great performance – but making so much in one move! These were apparently two entirely different ways of expressing life on Earth, and both have been followed up fully. If plants' produce is ultimately honey via the bee and milk via the cow, it shows in this alone clearly a remoteness in the plant world and there is no way to compare this with a human 'plant'.

Wildlife is wild life in the true sense when it comes to its manner of expressing in its sheer unlimited variety. Plants lead, as it were, a double life. They live for the sheer sake of existence in the first place, but secondly they are there to feed fauna. Life, in short, acted accordingly from the start. If a gleam of life reached Earth by means of a meteor four billion years ago, then that surely means there is life elsewhere, which only deepens the puzzle as to what originally started it all. In other words, the core of it all remains one huge mystery. There was no suggestion at the start that development would result in plants and animals, but then neither is a plant seed much of a giveaway! Plants are the surprise package in green, since green was to be much more that just a colour in plants' view (eyeless though they are). Due to that fact, plants are also much more self-contained than animals. Indeed they could perhaps 'go it alone', with wind pollination, the soil to feed their roots and the sun's light.

The power of a flower has to be much more potent than any-thing like it in the animal world. After all, a plant and its flower(s) is an immobile thing and that means attraction has to be at least double the strength of that in the kingdom of animals. Flowers are not just tuning in to the requirements of the opposite sex; they go much further and attract creatures all the way from the world of fauna to assist them. Plants, so to speak, saw the light, providing the way ahead of not just their own green development on land but, what is more, also that of the other world, that of animals. It turned out to be a matter of two living worlds all in one – on one – planet. A colourful flower is clear evidence there is much more for plants then being a living green image and this is starkly emphasised in a bee orchid. It is well worth having another hard look at the green living world. Flowers are so much more remarkable, because they used to be no more than a fantasy, until plants' sensing created them millions upon millions of years ago. It is in plants' case too where *lebensraum* meant a structure the size of Earth. It is also quite interesting to realise that so much developed in the green world from the theme of a leaf. The only obstacle left in a way is Earth itself, by not being bigger to allow more development. There must have been reasoning behind it all to create life on Earth, because plants alone are highly complex, but they had to materialise to make room for a totally different living world, that of animals. If it had been only plants that were earthbound, they would still have been very special, but that combination of plants and animals makes it so much the more remarkable. It gives starkly the impression that it was all pre-planned. It was to be a joint effort which turned out to be a great success. There have been no fatalities on a large scale within their worlds, except that of the dinosaurs, but that was perhaps an accident. Plants show awareness in at least several ways and it is their fabulous creation of the flower that stands out above all.

Other interesting forms of awareness by plants are in stinging nettles, expressing the burning effect, which translated into Dutch is *Brandnetel*. In both cases that dreadful effect is expressed twice over in one word! Thorns on roses and brambles make sure there is an acute sense of awareness present. Furthermore, if there is too much munching of caterpillars on leaves, then a tree sees fit to warn its neighbour close by of the unpleasant experience – and this second one will increase its acidity. Another striking example of awareness is mimicry, as expressed between a dead-nettle and a stinging nettle, bearing in mind the two are not even of the same family. It means a dead-nettle is not troubled by intruders. Awareness is also expressed in a flamboyant manner in a plant with a death trap for insects. The goodness is sucked out of the insects' bodies.

The immobility of plants makes them even more alert, with the result that they flower in most cases. The flower seems to have the upper hand in comparison to anything in animal fashion. The green world is responsible for a huge increase in wildlife. Awareness in both worlds, that of flora and fauna, came with the creation. Plants show more of it by their expressions, especially colourfully in their creation of the flower, as a result of being immobile. It consequently gives the stark impression there was an intelligent power behind it all. How else could plants have created flower? Without the potential known as intelligence, flowers would never have been created, let alone a species like us!

The very beginning of it all, which can't have been more than a germ, could have been compared to a certain extent with that of a plant seed, which is also crammed with data and results, in most cases, in a plant flower object. The potential of sense was there from the start. It led to the invention known as the assimilation process and later on, in the same unit, the creation of flowers.

Furthermore, the most vital importance of a root system is also a very enigmatic creation, and one which plants cannot live without. It leads to, among others, plants the size of trees. They are the giants in the plant world and express the quality in length, while that of a flower goes in depth. Well-developed flowers are three-dimensional. They attract by means of aroma, and also by means of goodness within, known as nectar, and last but not least they attract by their beautiful shape and colour. All those qualities in flowers are for one definite purpose, namely fertilisation. Plants need to tackle this most complicated process for survival. Photosynthesis is one great complexity.

For animals on the whole, it is much too far-fetched to be rooted to the spot. Plants had to make do with such an unlikely state of affairs and, urged on, they came up with clever ideas to come to terms with their immobility, quite successfully. They invented stings and thorns, and increased the amount of acidity in tree leaves if they were bothered by too many intruders, but their greatest ingenuity is of course flowers. Despite the fact that they are immobile they nevertheless spread worldwide, including trees in huge masses like rainforests and in a huge variety. It sounds too far-fetched, but they managed somehow! It is amazing that there is so much information about plant life, as shown in this book. It is their assimilation process and root system which resulted in a most unique kingdom.

What made it necessary to incorporate part of sunlight into the life of plants? That is the big question. For us, the most highly developed beings, it is beyond our knowledge. It concerns a complexity out of this world to serve as part of creation, yet it is part of the plant world. To be rooted to the spot throughout life is also a very enigmatic issue, and we can't imagine anything like that either. It makes plants perplexing as a creation. We developed over billions of years, from the beginning, via evolution, and even the middle ages are now way behind us in our progress. Leaves dropping in the autumn is a matter of super-sensitivity by trees as a pre-winter precaution.

The green world, despite its huge disability of immobility, grew into a gigantic worldwide kingdom, with plants growing in close conditions in many places. Once a species became rooted, its spread looked to be possible only in a very restricted manner. The creation of flowers has a great deal to do with involving the entire world. Although, as creations, plants and animals are worlds apart, plants nevertheless depend on animal (insect) activity to a great extent, and animals are entirely dependent on plants. Having said that, they originate from one life core all the same. The entire issue is exceptionally enigmatic. A creation as utterly complex as that of the plant was clear evidence that much more was to follow, with the ultimate creation being us, Homo sapiens. By far not all realise that greatness, alas!

If life on earth had only consisted of single-cell creatures it would still have been wonderful – in the true sense of the word. There was, however, much more to come, and it is expressed in three-dimensional features everywhere. There must have been an enormous force behind it all, however small the original germ might have been. Plants were the first to prove that creation went into depth, and we are the last on that list to prove it, though in a very different way indeed!

Flora is sufficiently interesting for the general public; there are programmes about it on radio and television. This, because plants are immobile, had to be laid on to make surviving possible despite that handicap. It all went by feel in a big way, and even attracting bees, among others, with their flowers. It must have been a huge undertaking at the beginning of life on Earth, to realise the assimilation process, which was a matter of involving sunlight in the actual process of living. It was also a matter of being interested in matter outside Earth. It makes the sun a second participant in it all. Flora's basic structure is paper-thin leaves as a rule and such a relatively small fleecy appearance also created plants in the form of sky-high trees!

It is amazing that the minute unit, life's beginning, managed to incorporate sun's light to survive. The assimilation process was most important from the beginning and responsible for the existence of all life on Earth. It is most extraordinary that life ever developed, since it is more remarkable than just being 'outlandish'! In other words, the assimilation process is an out of this world connection in a literal sense. It makes plants more than just a worldly creation. Life, at the very beginning, before it touched Earth, came about as a thought before writing. A house is just a fantasy at first, before brick and mortar turns it into reality.

That famous process was taken on board from the very beginning, which makes it more remarkable. Without it there wouldn't have been any creation, other than primitive life at best. It is also interesting to realise it is us who, at long last, after all the creativity over billions of years, managed to absorb the enormous value of it all, but we are still in the dark about its very beginning. Indeed, the start of it all must have evolved due to a power of unbelievable strength. It all happened in a 'just so' manner! We ourselves are the ultimate outcome, realising most of it, save for the very beginning.

The variety in paper-thin leaves is enormous, including the variegated ones. It is itself a clear sign of the gigantic impact creation has had on it all.

The containers of nuts would become coffins if they weren't sufficiently porous for the compressed life within to breathe. It is again a structure, like an apple, with transport in mind. The only difference lies in the appearance. It shows here too quite clearly, how much value is attached to ensure that the goods are delivered. The message, in other words, in both cases is quite plain that way. An enormous amount of evolution work went into creating both apple and nut.

*Vive la difference* was the motto from the start, roughly four billion years ago – and that makes it a long cry. Differences increased in proportion with growth. A pattern of output eventually developed, better known under the combined name of plant and animal, truly a world concern. In other words, a joint effort of the first order.

It is the more perplexing that a process as complicated as that of assimilation materialised at a time when life on earth was still very primitive. It is a feature which is of such a high standard that it is beyond us. How could an extremely primitive species, as it was at the time, possibly have developed such a highly complex unit which was very necessary indeed to spread life on earth. There must have been a super-power present in it all as long ago as about four billion years. It meant a truly worldwide spread of flora and fauna all in good time. Without that incredible invention, life on Earth would have been very restricted and equally primitive – if anything at all! We are the lucky ones to be able to take it all in, but for the working of that famous process and the very beginning itself.

There is a most interesting second way of behaviour in the plant world, which again makes them a very special living world, and that is their extraordinary root system, which is absolutely massive in the lives of trees.

We owe it indirectly to plants that the world of animals suffi-ciently developed to lead to us after almost four billion years. No wonder it was considered to be important enough to write a book about it, being mainly a philosophical look at plants. The creation of plants was clear evidence that life on Earth was going to be great, three-dimensionally.

Creation of plants is such a complex matter; nothing can be compared with it – not even us. After all, we too developed by 'mere' evolution over billions of years!

Plants had to be on the scene at about the same time, however primitive at first, as when animals started to develop, since the latter were depending on the former to be food suppliers. Plants, in other words, control life on Earth. When plants came into existence, it was a matter of realising the unbelievable. It came out of the blue or, in other words, was an out of this world issue – of which we see the evidence all around us outside. It has always been a matter of making the best of it, which concerns the very restricted area indeed in plants' case – a seed dropping spot – and it necessitates them probing all possible potentials available in such a zone. It even came to an eruption in trees, reaching for the sky. Animals created features outside their own unit, like birds' nests, ants' heaps, beavers' dams, bees' hives, and it went from there to much more in us. In plants, the 'office' had to be within, since travel was not possible for them. Home ground has a much more serious ring to it in plants' case. They rely on it entirely. Only plant seeds realise travel, indirectly.

Life is the biggest show on earth. We are all in it. The question is: who wrote the script? Clearly there is more to it than it just having happened in a purely accidental way. Mimicry is scientifically certified as a phenomenon – or rather, the phenomenal part of life is surfacing in that issue. That life in general is still everywhere on Earth after billions of years is in itself most amazing. The more so since a single life is not only very delicate usually, but doesn't last very long, a matter of decades in many cases or often even less. It turns the whole issue into a most perplexing affair. The creation of plants billions of years ago was clear evidence that life on earth was going to be enormous in dimensions. We see the evidence of it at present, although we cause a decline to quite an extent in both worlds, alas. The invention of plants was bound to lead up to such a greatness and it was not just a matter of microscopic organisms spreading over this planet, which all the same would have been most remarkable in itself. It developed an awe-inspiring quality. What impulse made life ready for the photosynthetic process? It was a very big undertaking and all life on Earth was going to depend on it.

To go into detail for a moment, there are grubs at the pre-imago stage, which spend no less than a decade or more underground (one American cycad mines no less than seventeen years underground and turns into a fully fledged insect for no more than a few weeks). That's not fair, we are inclined to think, and one wonders why they go through all that trouble. When the grub, like a caterpillar, turns one day into a most exquisite butterfly or moth, it shows there are dimensions involved, to evolve to a much higher standard than 'ordinary evolution' could possibly have mastered.

To flourish has been expressed *in optima forma* in flowers. Flowers concern the most exciting wildlife story, namely where two worlds meet, in which case the plant is the objective element and the animal (bees among others) subjective. Many features of awareness in plants, including their flowers, show that making sense is as much a part of their way of life as it is in the other kingdom. Leaves turning edgewise towards the sun during intense tropical heat, avoiding scorching that way, is only one of numerous examples.

Plants, extremely complicated as they are, had to be realised at the beginning to enable life to evolve.

The potentials were there in a huge magnitude once leaves were created. They resulted in two living worlds directly and indirectly. Life's interference with Earth, however minute at first, was going to be a world all by itself. One of those huge sectors of the living world, that of plants, created yet another, that of flowers, besides a totally different community which is also colossal in size, that of animals. In other words, the potentials from the start, once the assimilation process was put into action, were absolutely enormous. Plants caused a breakthrough, enabling a second living world to develop. Plants' establishment on earth was very strange indeed to say the least, indicating that there was a lot of go in the original life on Earth. In animals' case it is a matter of, 'have legs, will move', while it is exactly the opposite in plants' world, being, 'have roots, will stay put'.

Life was soon to follow a path of enormous expansion, once it had got in touch with this globe. More, it did so in a two-fold way, which in itself is most remarkable, and only possible due to the highly complicated assimilation process.

In most plants, colour is much more than an aesthetic value, as it plays an important part in the fertilisation process. The interesting thing is that sophisticated plants consist of two completely different stations, of which one, the green one, focuses on the sun, and the other has animals (usually insects) in mind.

We owe our being here to sunlight. Without the assimilation process, animal life couldn't have evolved. Animal and plant life came about by sensing, and a plant's flower is an outstanding example of this. Then there is us – Homo sapiens. We were created consequently as the last and greatest performance, save for some mishaps on our part.

Sensing, expressed in sensitivity, shows throughout life in both worlds, those of plants and animals. It shows everywhere in our own world, alas not always in a positive manner. Life went from abstract into matter and plants still have an abstract basis.

It is an expression of the abstract turned into matter, however unbelievable this may sound in plants. An identical process may have been present, from abstract into matter, in the case of the life core. The feature of abstract is at least part of it all, and there again evolution is abstract-based. We see the evidence all around us in the countryside as far as abstract is concerned, shown by green matter. Life on Earth was not just an odd coincidence, but must have been a serious, pre-planned feature, for it all fitted too much in a precise manner. Plants were a necessity as well as being a fantastic living world in their own right. Plants have it in their power either to attract or repulse creatures. It is all under their control. Their systems had to be foolproof, being their only option as a result of being stationary. To be immobile was just asking for trouble unless extraordinary ways were found to overcome this handicap. Any plant (or tree for that matter) is a sensitive plant. Thus the flower is a display of sensitivity. It is a focal point of a great deal. After all, bees for one have been made to pay them a flying visit. A flower is where sense turned colourful.

Expressiveness had to be so much more pronounced in plants, due to their lack of moving in the true sense. That's why it led to roots, to flowers, to huge growth into trees, to branching out, to fruits and to plants' eggs – the seeds of which many may travel by air.

A flower is where a form of telecommunication has been realised. There is a 'hotline' there with bees (and others) at the other end. This 'linking up' issue had to be a very striking one but is yet perplexing – however much we take flowers for granted. The two specific areas of growth in plants, the leaves and the flower, make a plant only half an Earth dweller and partly a Sun creation. They stress the importance of the sun and its light in a most emphatic way. They express the sun's light as much more than just light, in a manner of speaking. There is more to be had out of light, according to plants!

Every year in our occidental regions, we witness one of the most spectacular events, namely the time when all the leaves of many trees drop down – or, as the French have it: *Quand les feuilles tombent*. It is a matter of sensitivity in the extreme and a preparation for the harsh conditions that will follow months later. It is a *plan de campagne* realised literally by those plants known as deciduous trees. A kind of instinct displayed by the green world, it is a premonition of the green kind, showing plants sense things too.

Plants were the most important creation, one that animal life was to depend on. Plants' territory is not only definitely for life, they penetrate terra firma. Plants go through endurance tests animals couldn't possibly face up to. They are weatherproof by any standard and animals envy this. Endurance tests are even stressed that way, where flowers develop stalks. Although a stem is very helpful, on the other hand it increases exposure dramatically. Many are in a way as much 'sun flowers' as the true sunflower. The latter is a giant among flowers and flowering plants. Sun's part in it all is indeed more than Earth's share, and that makes the sun quite big in another much more profound way. Plants managed not just to survive, but even to blossom. They made it all work by means of an amalgamation and absorb light into their units on the one hand, and usually attract animal life to the most precious part of their works, the flower, on the other hand – which is just another way of 'absorbing'. One could say plants saw the light and never looked back, progressing onwards from then.

If there is life on another planet, then there almost certainly always has to be a green world first and foremost. Then it might be possible to find beings there as well, with any luck intellectual ones, besides many animal-like creatures.

Have legs, will travel is the animals' way. Plants on the other hand are as stationary as they must have been in the beginning. What made plants develop roots is as mysterious as the process of photosynthesis. Plants are supernatural in their assimilation process and as a basis of all life on Earth.

The birth of a plant seed is as phenomenal as a seed developing in a tree. The actual first germ of it all could well have been infinitesimally small. Plants and animals have nothing in common other than creation, and therefore they probably started life separately.

Even today, there are insects so small one needs a magnifying glass to see them properly.

Plants also show themselves to possess a spirit for survival's sake. This resulted in plant extremes galore, like massive tree trunks and paper-thin leaves (thin enough for pages of books to be called leaves in Dutch and Latin) or issues like living underground and above ground simultaneously. Furthermore there is the upright style of living in plants (especially in the case of trees) as well as hanging down or hanging on. Others do it the wriggling, twisting, stretching or even the running way. Attracting indeed goes a very long way. It not only reaches the bee from afar (as an alien!), it also made flowers. They are all bits of a big jigsaw puzzle – and obviously, for us, quite a few pieces are missing! Wildlife is feeling or sense embodied. Sense certainly did not turn up halfway, but was there from the start.

Life found a base on Earth due to homing-in, like swallows arriving home from thousands of miles away, or a dog finding its way back home over an incredibly long distance.

To produce seems like an out of this world concept. It still sounds too far-fetched to express in words, yet is of vital importance to life on Earth.

Think of a large, lone tree in a field and it may seem incomprehensible to put it under the kingdom of plants. A beautiful wild flower is the ultimate in creation of wildlife. It is nature's greatest performance: a plant seen at its best. Plants found the strength, due to supernatural living, to penetrate in the ground they stand on, in a process known as rooting. To involve the sun's light in living, as in the plants' case, is too far-fetched for words to express. Flowers are sense blossoming. The big question is: what force lay behind it to use part of sunlight, of all things, to continue life? There is a might behind it all of which we see the evidence, but it is beyond us to measure.

This sense of feel not only made a plant flower for a very good reason, but also enables us to see it as beautiful.

That plants had come into being as primitive units, in tree ferns and horsetail plants, was in itself already a refined piece of work. However, evolution went on, reaching the stage of a flower. It turned out to be, in most cases, a unit almost entirely its own and at least as complex as a green plant. The only slender link between the two departments is a stalk. The outcome has a far-reaching effect, emphasised by attraction. A plant has much 'go' in itself, without so much as making a move.

The assimilation process proves the point that life was going to be a very serious affair, to an almost unbelievable magnitude. The latter shows it in length, as spread, while that famous process expresses it in depth, usually in paradoxically paper-thin leaves. The start was a beginning of a dual world spread in quantity, but above all in quality. Plants show there is a process from abstract to matter, which were very close to each other at the beginning of it all. Maybe a similar state of affairs from abstract to matter might have been at the start of it all, however minuscule in dimensions. This base, Earth, is in touch with the sun. The latter was going to play a major part in creation. It turned out to be an enigma realised. Feeding on vegetative matter is as important as breathing in air for all fauna.

It is in plants' roots where the organic meets the inorganic. Nowhere else in life do the two get closer, and it is yet another very striking feature in plant life. It is an amalgamation of a very interesting kind. Life is shown, especially in plants, to be unable to get away entirely from the inorganic world! It is underground where the two get close, yet are worlds apart at the same time. The bond between the two is only seemingly one and it makes life a realm entirely of its own, in which both Sun and Earth are merely playing the role of subjects. Had Earth been thousands of times bigger, intellectual life might then have been so much more interesting still – if it had only been a matter of how to reach each other from end to end. It is all a matter of *lebensraum*, and we find that now quite a hard one to come to terms with. To be *sans famile* in the universe sounds most improbable. Yet it is so, any way – until we might know better.

It is also a penetration in root fashion, which is of actual importance – not just to plant life, but to us all. It makes us all 'anchored' to Earth, however much we fly away from it nowadays. It is all a case of penetration, turning mass into biomass. Plants show this quite clearly. They exist by using sunlight, producing via photosynthesis their chlorophyll. Plants are 'plans in green' which paid off. The same goes for the animal world.

It turned out to be a truly silent explosion – or a completely noiseless big bang. The great expansion, in a two-fold way, reached eventually an ultimate goal – us – as an end product. We ourselves are the personification of the unlimitedness of it all – limited as we may be. Our display at large today would have been considered impossible by people living during the Dark Ages already, let alone in the 'darker ages' still, those of the cavemen – and then there were monkeys before that – and so on!

Plants are invaders in the sense that they penetrate the ground they stand on, as well as absorbing the light above ground for their survival. There is an 'upstairs/downstairs' image apparent in their way of living. Without the green world, there simply wouldn't have been life on Earth. No wonder it went to such lengths in the form of the highly complex assimilation process to realise it. Well-developed plants are enigmatic in three different ways, namely that mysterious process of photosynthesis, their rooting capability and their creation of colourful flowers. Plants show that creation was an enormous force (and still is), capable of absorbing the sun's light. Plants are a creation we cannot fathom. That they ever came into being is as far-fetched as a proper fantasy. They are down to earth literally and figuratively on the one hand, and from another point of view truly out of this world bearing in mind their connection with the sun's light. We owe our lives to plants.

This book was realised by abstract thoughts becoming matter, by putting them down on paper. Not unlike, from that point of view, abstract light turning into green matter. Such a huge effect as plants express shows that this planet was ideal for an entire living world. Plants are an explosion in green, realised by the mysterious assimilation process. They are an 'outlandish' family, seeing their bond with the other heavenly body, the sun.

The creation of a green leaf is so thin that a page is actually called a *bladzyde* (leaf side) in Dutch. The creation of a green plant leaf is in a way more complicated than the actual arrival of a life germ on Earth. They are both mysterious. The latter, however, was crammed with data, like a seed. Plants were the first evidence that original life on Earth was going to develop on a colossal scale. To decipher the phenomenon of life itself is alas impossible, but a great deal of data can be gathered about its surroundings. Plants for one display a process of which we know the outcome, and that is most awe-inspiring. So is a beautiful wild flower, a very grandiose feature of wonder, as a mysterious creation established in the plant world.

That which made contact with this planet was actually the very beginning of a germ of life and not at a halfway stage. It is therefore obvious that it was all pre-planned and the assimilation process made that so much the clearer. One highly complex process, that of photosynthesis, was going to bring about two gigantic living worlds, spread over the entire planet of Earth. The start seems to have happened on this planet, rather than arriving by meteor. There may have been many untold meteors crashing on Earth over billions of years, but none contained life, let alone the very beginning of it. The life core may have been very minute and inconspicuous, but what an outcome! Plants themselves are in a way a research station: a probing on the spot, a laboratory within, where chlorophyll is produced by even the humblest of plants. It is a knowledge still baffling us today and making us still depend entirely on them.

Plants are proof of life's intention to develop here on Earth – and plants in fact do so much more than just braving the elements all of their lives. The latter alone require almost unbelievable endurance, and it sets them apart from the animals. It is their *piece de resistance*. Plants, once created, showed what life was made of. It was totally unexpected. It is a fabulous manifestation as an image. However complicated they may be as a creation, they had to be introduced to realise a second living world. It is amazing to say the least that a seemingly simple life core visualised a creation as highly complex as that of plants, and relatively soon after its contact with this planet too. In other words, whatever came in touch with this planet had huge potential. Plants were the first evidence that something big had got in touch with this planet. Its dimensions were partly out of this world, as soon became clear. It is most likely that the actual creation of an infinitesimally small unit took place here at a convenient spot.

The major part of a planet's creation is abstract-based and life may have started in a way in which abstract played a major role.

There had to be a huge display soon after the beginning of it all, to realise life on Earth in its colossal dimensions – and that was the creation of plants, which involved a manifestation out of this world, namely part of the sun's light. This was the biggest performance in the creation of life on Earth in which an alien might was introduced. Then there is the rooting system, which in some places is strong enough to lift up tarmac or big stones. It is amazing that a creation as far-fetched as plants not only settled on this planet, but in a huge extravaganza and worldwide.

It all went by direction, or how else could something so weird as using light in a very complex process to enable plants to survive have the capability to expand life on earth? It is the more striking since it happened relatively soon after contact was made with this planet. The environments of Sun and Earth and the life core being capable of enormous expansion (partly due to the sun) make it plain it had to happen. It is plants again, which are so clearly down to earth in their rooting system on the one hand yet out of this world on the other hand with their photosynthesis process. It makes life on Earth very special indeed!

Instead of development in the slightest manner of a simple life form over billions of years, it expanded in quantity and quality on an enormous scale, due to the arrival of enigmatic plant life. It led ultimately to us, Homo sapiens.

Once the sun's light got involved in the living process, it showed that life on Earth was going to be a huge undertaking.

The sensitive plant shows only more clearly how sensitive it really is, by folding up its leaves when touched. It demonstrates this trait also quite clearly in the dead-nettle copying the stinging nettle, but for its stings, and in a glamorous way in the bee orchid. It is this sensitivity, as abstract matter, which was present from the very beginning. It made flowers among others in principle – indeed, it made plants in the first place. It shows that abstract matter played a major role in it all, including the beginning itself. Plants were the first tangible evidence that life on Earth was going to be huge in dimension, in a worldwide fashion. The most remarkable part is that it ever came to plants!

Plants reproduce by means of seeds, either by wind, by ejection, or travelling by bird etc. Space, in this, is very important. Great care has been taken to see that seeds do not dissolve in birds' stomachs. Seeds will only germinate once they are dropped off as bird droppings; most unglamorous! It borders on the unbelievable. It is all plants' way of looking ahead, headless though they are.

The plant had to be invented before creation developed other species by evolution. The creation of a bird is explainable, although only just, and so is the creation up to Homo sapiens, but the creation of leaves defies any explanation.

The most unlikely thing happened when part of the sun's light was co-opted by plants for their existence.

Mimicry is a way of expressing life in a mirror form, and it happens in animal as well as plant life. It shows the incredible strength of sensitivity – for a very good reason! Plants are in the closest possible contact with the sun, the soil and the elements. They live within that framework. It is interesting to realise that plants, which have so much 'say' in life on Earth, are in fact the utterly silent ones.

Plants came into being in the most unlikely way. To use sunlight for existence would have sounded too absurd for us, had we been around at the beginning. In fact, as it is, we depend on it!

Leaf green is life green. The actual life-core must have been most special in its own way, with the ability to involve part of the sun's light in the creation of life. The substance responsible for it all is chlorophyll. Green leaves are consequently the closest link between Sun and Earth. It is in those usually paper-thin leaves that abstract turns into mass. Light and life are, in other words, in very close touch with each other. That goes especially for the plant sector, where leaf green is much more than just a colour. From algae to flowering plant, reaching bees and others in the process resulted in an immense magnitude. Plants made it all happen. Well-developed plants show in a three-fold way an extraordinary existence, namely their underground rooting system, their speciality in flowers and above all their out of this world assimilation process. It makes them a world all of their own. Plants, as a green option, were the only way to develop a necessary green world on this planet. It was the life core which was responsible for creating two living worlds – worlds apart in their structure.

Plants have a 'field day' every day. They are 'outdoors people' *par excellence*. Looking for shelter is not possible for them. They simply die off in many cases, as annuals, almost before true winter sets in. In the case of perennials, a small remainder spend life underground during the winter months, as hardly more than a biomass of roots. Going underground is for such perennials a matter of survival, avoiding in that way contact with the arch-enemy, winter. It is totally unknown in the other kingdom. It is a matter of coming to terms with reality of a completely different kind. However, it is their only option to stay alive – or all perennials would degrade to annuals. Such plants spend most of their time inside seed-capsules, not exactly satisfying. It is drastic in the extreme.

There are creatures for whom plants have almost become an extension of their lives, like a squirrel living in a tree, eating its nuts, resting and nesting there too. It goes to such an extent that it is almost a rarity to see one on the ground. For caterpillars too, day and night they munch plant leaves, often in an almost fanatical way, until they become butterflies. A sloe in the tropics becomes almost part of a branch itself, and it is totally helpless without its support.

It has turned in such cases into an amalgamation *par excellence* or in other words an inseparable union. Plants, once they came into being, were evidence that life was going to be spread all over the globe. They were not just an idle attempt to try something different instead of very primitive life on this planet. They were a clear expression of a proper start in life on a huge scale. Plants show that abstract matter can be turned into biomass, which in their case is leaf green and quite possibly the creation of the actual core came into being in a similar way, namely abstract into biomass. It is enigma turned into reality. Plants show it can be done, however mysteriously. It is like a thought turned into matter – or from will to power.

It is amazing that primitive life, as it was there and then, could possibly have been attracted to something so utterly complicated as leaf green. We wouldn't have been here if it hadn't done so, it is that important. This 'reasoning' started no less than four billion years ago. It is leaf green resulting in trees; it is leaf green which results in carpeting many places as fields on Earth; it is again leaf green which makes plants flower, attracting creatures from the other world; and it is yet again leaf green which makes crops like corn grow.

A plant is raw material for the creation of a flower, from which offspring is eventually created.

Plants give us beverages other than plain water, which is only meant to quench our thirst. Orange juice is a well-known one, and we must also include 'firewater', or alcohol, which is not always good for us.

The sunlight itself, which would eventually contribute greatly to the creation of plants that we know today, has been picked out by means of ultra-sensitivity. It opened the sluice gates, so to speak, not just for the spread of plants, but also for the animal kingdom. That famous absorbing quality gave rise to that all-important photosynthetic process. Plants, in other words, did a good deal more than merely looking for the sun (as is so often overdone by humans nowadays), but actually got in touch with its rays, simply 'to be' – without eyes to see (which wouldn't have done them much good at any rate, and would only have blinded them in this case). In summary, sense led to a photosynthetic process. Sense and life are at least very closely in touch with each other – almost to the point of being one and the same – not unlike plant and flower.

A tree is a plant as much as the grass at its base or a bee orchid. Their variety on that theme is at least as large as in the animal kingdom.

Chlorophyll is responsible for all land life on Earth. Life made contact with this planet, because it was all set for it, with its climates and it is still going strong after four billion years, delicate as the living world may be. Chlorophyll was out of this world to create (and still functions today), and was all-important. Plants were formed at the beginning and evolved later in a huge spread of all sorts, while animals came about 'merely' by evolution.

Plants started life due to a highly complex process and then developed by means of evolution. Animals on the other hand developed very slowly, from almost point zero. It gives strongly the impression that everything went 'according to plan'.

Plants penetrated into deepest matter, including the abstract and the roots job as so to speak peanuts in comparison to what is actually achieved well above ground – by the same unit.

Sense is the 'traveller' towards potentials in the universe – of which Earth is but one 'home'. Earth is where life could get itself rooted, literally in plants and even in us in a metaphorical sense. We represent 'a penetration into matter' *par excellence*, by far outdoing roots in this.

Abstract (sense) turned into green matter in plants and grey matter (brains) in many animals. It is realised on a modest scale in mimicry.

Plants, not including their flowers, are the independent ones and get their own sustenance from the soil and the sun, whereas animals rely totally on plants. Plants often feign death, copying it in a way almost like mimicry, by dying off in winter – or apparently turning into a skeleton. It is for them however (trees, usually), a symbol of being alive in some way. It is most interesting to realise that a type of very complicated plant, with its out of this world assimilation process, is the only-too-well-known grass, in fields and lawns all over the world. If only we gave a moment's thought how downright precious our lawn really is. To involve part of sunlight in the actual process of living goes beyond words to express it. It is such a super-might which was apparently necessary to realise life on Earth.

This green matter enabled us to be the optimum of all living. Plants incorporated soil in their systems, which led to perplexing developments and the creation of. Plants' structures, especially manifested in a big way in trees, are the outcome of that enigmatic substance chlorophyll, which originated from the abstract. It is obvious that life made contact with Earth, because of the climate. This planet was ready for a world expansion and the sun was a cooperating might. It is also obvious chlorophyll was created because it was the only option for a worldwide spread of life. Due to it, we are here. It was going to be a silent explosion or a worldwide spread once life made contact with Earth.

A plant appears almost entirely by abstract means, using chlorophyll, and a cutting turning again into a fully-grown plant, shrub or indeed a tree, is also bordering on the unbelievable. Both the creation of chlorophyll and a cutting growing demonstrate that a very small start leads to a gigantic happening. It made sense – the life core created it! At least one of the two kingdoms has almost entirely abstract (light) as a basis.

Plants are for animals downright weird, and of course the reverse is true also. There is cause and effect in it all. It is even underlined, and there is a 'cause' (plants) and an 'effect' (animals) reaction. Many plants come to an almost complete standstill twice over, namely by means of roots, which cause complete immobility, and, when winter comes, perennials don't just go into a deep sleep but achieve their goal to overcome the cold in a much more simplified manner. They appear to die off, which couldn't be closer to the bottom of the matter, meaning reaching point absolute zero – almost!

Plants have the ability to attract creatures from the other world to their flowers for fertilisation. Plants in that case are like aliens and the point is stressed since they absorb the sun's light for survival, attracting creatures from another world to their flowers. Animals have developed due to evolution over aeons, but the plant had to be created first and it even had to involve another heavenly body to do so: the sun. That was necessary at the beginning before evolution was to be the abstract force behind it all in realising a huge variety of plant life too. Plants are the basis. They are fundamental to animal life. They were the most unlikely happening, yet an absolute necessity for animals. It gives therefore the stronger impression it was all pre-planned. Plants are biomass that came about in an enigmatic way, originally. They are from that point of view as far-fetched in creation as chlorophyll itself. So much was going to happen once plants came into their own. It resulted in an entirely different living world. Plants are abstract-related, not unlike the beginning. Plants and animals contrast in the extreme. The former gets rooted on the spot, while an animal's priority is to move about, to mention just one difference.

This philosophical look at plants gives clearly the impression it is well worth taking the green world more seriously – after all, it resulted in us, Homo sapiens. Plants are a chlorophyll-rich creation, realised indirectly by the sun's involvement.

Trees have might in height, while flowers show it in depth. Trees express in their way the enormous viability in plants, by far outstretching growth in the animal kingdom. They show the might involved which created the plant world.

Mature green leaves show the basis of life is abstract to a very great extent. Trees are green biomass in the extreme. Plants go to extremes as far as survival is concerned and perennials simply die off above ground and carry on living underground. This is a way of life totally unknown in the animal kingdom. There are extremes galore in plant life, like the way trees expand, and then there are the well-developed flowering kind of plants, which demonstrate their extremes by attracting creatures all the way from the other world. Furthermore, there is their rooting system with the extreme of going deep underground. It is all because of the most extraordinary creation, expressed in their assimilation process. One more extreme is the fact that deciduous trees spend winter time as virtual skeletons. Plants prove that despite having no heart, brain or body, much can be achieved all the same. They still spread worldwide in a huge variety.

It was a chance in a million that they ever came into existence; for that, the assimilation process seems too far-fetched. Plants are enigmatic biomass, which all life depends on. A plant leaf consists of two sheets, and a leaf miner, which is a tiny creature, proves the point by going between the two layers to make a living. The plant world would have been severely restricted in its development without flowers attracting creatures from the world of fauna. A plant consists of 'bits' put together, i.e. the green life, roots, flowers in most cases and, lastly, the assistance from outside, which is of vital importance (involving the fertilisation process) to their unit.

A bee is an alien to a plant – which is not far off our idea of an alien visiting another planet!

Plants are the outcomes of two poles, as it were, with roots on the one hand functioning as a pumping station and leaves absorbing light from above at the other end.

Light to chlorophyll is as mysterious a step undertaken as the 'doubles' in nature (that is, mimicry). It shows in a nutshell the concept of a plant as a go-between, from abstract to matter, not unlike the sun, via plants and animals, which has had at least a great deal to do with materialising all land life.

There are birds barely touching terra firma any longer. Abstract sunlight was going to be the basis for plant life and it still is an important part of their daily sustenance. Plants, in other words, depend on the assimilation process, the weather and the soil they stand on, while animals just depend on the weather. Furthermore, plants' main motif is a paper-thin leaf, usually, whereas animals have bodies. Those paper-thin leaves led among other things to massive trees and flowers.

The sun is involved in the life process of both worlds. It may be a matter of indirect involvement in the case of animals, but that doesn't make it less important and proves the point that two bodies are involved in life on Earth. One might conclude that there would be a similar situation anywhere else in the universe where there is life.

Plants also feed themselves on organic matter in the soil (besides processed abstract matter, that of sunlight) that was once plant or tree material.

It gives strongly the impression that a protozoa as the first representative of wildlife on earth has been remote-controlled to reach ultimately the stage of us, with the incorporation of plants as a necessity. In other words, intellectual life here on Earth could only have developed with green life, that of plants, as basis. Plants also create plants, by means of a cutting stuck in the ground. It is just another superlative way that plants express life. It turned from abstract into green matter, creating plants that way. It was to be the basis of all life on Earth.

Once a life unit was established here, plants had to be created to assure a two-fold worldwide spread. The abstract has a major share in the creation of life on Earth, especially in plants, which are realised from abstract matter, processed by chlorophyll and they are of first importance to all life. These two, sunlight and sense, are abstracts, so dear to life, especially in us, falling apart again to absolute nothingness in death eventually. Chlorophyll leads to leaves that are the fuel for fauna. In summary, green and sense together are to a very great extent responsible for life. Plants show the importance of telecommunications, being clearly 'receiving stations' themselves. Birds on the other hand show that the sky can be mastered with a body heavier than air – and humans have elaborated on that one quite considerably.

It was the only way, not only to expand hugely but, what's more, to achieve a pure amount of living – us. If anything makes sense, that does (though the issue may be blurred sometimes nowadays).

The photosynthetic processing is a plant performing like a chemical plant. We all owe our lives to this process. In other words, plants are totally lost without chlorophyll – and so are we.

There are two approaches for living, of which that of the animals is the logical one and that of the plants the far-fetched one. Who would have thought of incorporating a light wavelength within one's actual life! It all started from one given point and the big split of plants and animals as a result is an outstanding feature to say the least.

Leaf green is a substance of enormous complexity and abstract matter plays a big role in it, which is all processed and transformed into green biomass. Of all the creations, even including ourselves, this is perhaps the biggest performance of them all.

Life on Earth is almost entirely a mind over matter issue. The big question is: whose mind? The universe and life go hand in hand in this and we all appear to be subject to a huge mastermind. It makes it not so much a matter of life on Earth, but rather a mind over matter issue on this planet. Mimicry puts, as it were, a finer point on it and is therefore to be considered a VIP (Very Important Part) in the living world. Mimicry shows very clearly what life is made of in principle, in which abstract plays a huge role. It is not so much the clever copying part but rather the mind over matter issue showing through.

There must have been, indeed, a mastermind behind it all because it is simply too far-fetched to use sunlight for life purely by chance. It is because of that process that we arrived! Once plants came into being, there was the possibility of realising a completely different living world, namely that of animals.

It is sensing again, in a superlative, sensitive way, which created plants and consequently animals. The only way to live in plants' case was to involve a duo-chemical system via the assimilation process and in rooting, each functioning independently, all because they have no mouths with which to feed. The interesting part is obviously that all systems work on that dual principle; plants look after themselves and animals are completely dependent on plants on the other hand. It is issues like the one above which make wildlife so interesting. It has fascinated and intrigued me from a very young age onwards, growing up in a family with nature much more in mind than petrol! It boils down to trying to find out what's going on out there, like: what is actually a leaf, or a pin prick spider and yet provided with eight legs all the same, as if there is nothing to it – and so on. The issue which surfaces above all is the will to live. It often goes beyond expectations, but then, all life is one big surprise.

Flowers look almost out of place and a long stalk only increases that distance. The concept of a flower seems weird enough and even in one's wildest fantasies it will never go together with a fern.

Everlasting life is a contradiction in terms. It is more likely that life is very close to being abstract itself, since one of its major points – which so much life depends on – is abstract, namely wavelength green. Take away the sun's light and all land life dies. In other words: as long as there is abstract, there is life. Abstracts, like sense and the colour green, turn objective whereas life shows itself to be subjective in this. In death we turn abstract again, both fauna and flora. The plant had to be created, however unlikely, before other creations could follow by evolution.

Plants are independent apart from most of the flowering ones, save for their soil involvement, and it could have been a world of plants only on this planet, without the complicated flowering ones. The plant was created, regardless of its involvement with animals in mind. The power in life is overwhelming and more than just earthly.

Plants control animals through invitations sent out (via flowers) and the 'not welcome' notices, like those given by stinging nettles.

Caterpillars may not be exactly plants' friends, but a very

strong bond is there all the same – at least from the insect's side. It has to be heather for the emperor moth, carrot leaves, wild or cultivated, for the swallowtail, and nettles for the tortoiseshell and many others. That's attraction at its best, guiding butterflies and many others to that extent. It becomes the more impressive, realising we have to put heather, carrot or nettle leaves to our nose to smell the difference, while said insects come from afar, picking out the exact plant, solely for the sake of living.

Absorbing sunlight, attracting bees and others with their nectar for one purpose, attracting others with their aromas for another purpose and, last but not least, making it very plain when they wish to be left alone, again makes plants sensing stations of the first order. The essential part of a plant is, therefore, like a radio station, with messages sent out as well as received and understood, with the peculiarity, however, that the 'station' itself materialised accordingly. To be a plant, therefore, is to 'think abstract' to a very great extent. After all, absorbing, sucking up and attracting are all closely related, in which attraction brought about flowers, absorbing created leaves and sucking up resulted in roots. In other words, there is sensing galore. It also gave rise to mimicry. Plants also brought colours to life in a very flowery way via light to chlorophyll.

Plants are the immobile ones *in optima forma*, yet somehow they too managed to spread worldwide, which in their case is a most extraordinary performance. It is an expression beyond words.

Plants materialised due to an out of this world involvement. The creation of it was the first evidence of a super-power present, which later on was expressed in a worldwide spread. Flowers have been the result of an enormous increase in flora all over the world. It is sense again, in a big way, which put plants up to creating flowers. Both plants and flowers were created with abstract playing a major role. Plants are also strongly climate-dependent, which is yet again another abstract. They are in touch with it day and night, on the spot. Plants result in footpaths in the countryside all over the globe. It expresses the density in which plants grow everywhere.

It was super-sensitivity, again, that led to the involvement of the sun's light to create plants. It is also sense that played a major role in it all from the very beginning up to life today. Mimicry is sense expressed three-dimensionally. The life core is as mysterious a creation as the way plants came to life. In the latter case, (abstract) sunlight plays a major role.

The journey through space by seeds is extraordinary by any standard – the more so realising they are immobile plants' 'brainchild'. There are even cases where the travelling has to take place within a bird's body to render the seed's viability (through enzymes).

This volume reads like an outsized love letter of a different kind, and expresses a deep empathy for nature, especially concerning plants' role in it all.

Plants contradict animals. Where the latter grow thick fur in wintertime, look for shelter or even go into a deep sleep in many cases, escaping the cold in three different ways, plants, especially trees, have to face up to any hardship. They have to stand there come what may – naked, in the case of deciduous trees. Soil and plant create a bond which only brute force can break, causing an uprooting effect. A root gets closer down into the earth than even an earthworm, yet at the same time, soil appears as lifeless as a plant is full of life.

The creation of plants and that of the life core are both enigmas, and it may be true to say that abstract plays a major role in each case.

To orientate on part of sunlight as in plants' case for the sake of living is as unlikely as finding Earth a suitable globe for living. Plants developed life on Earth in a two-fold way.

The plant is where abstract (that of sunlight), via the assimilation process, turned into biomass worldwide in flora and fauna. In other words, it is this abstract which realised both living worlds, in an indirect sense in the case of animals. It shows how important abstract is in the creation of life on Earth.

It is worth going into this a good deal deeper since sensitivity in the extreme is expressed everywhere. It is, for example, in the case of a swallow most extraordinary that it finds back its way back to its nesting place after travelling thousands of miles, or a dog smelling some leaves at a specific spot on a footpath to find out whether another dog has been there before. It is also most outstanding that a green moth finds its resting place on green moss on a tree bark, making it nearly invisible. In the plant world there are also excellent examples of this form of abstract and the bee orchid is an outstanding example. Evolution in the plant or animal world is in itself a matter of sensing, and it created both worlds in this abstract way. Plants, in other words, are twice over abstract-influenced, namely in their actual creation, due to the assimilation process, and their development worldwide due to evolution. If the outcome was to be a huge world display, known as life on Earth, in which abstract played a major role, then so the origin must have been identical to that effect. Attraction, as an abstract, results in making contact, which is the way the life core met Earth.

It is amazing that so little of the sun's light, being processed, came to so much in plants' case, namely a worldwide spread. More, that part of the sun's light turned out to be of vital importance to the overall living process and to such an extent that we wouldn't be here if it hadn't happened.

It is the most perplexing issue that actual sunlight should be processed to serve as a green basis for all life on Earth.

It is by far the most outstanding feature of the entire living world. It makes it worth taking a second hard look at just a green leaf. Sun rays not only open up flowers, they made them in the first place.

In mimicry surfaces an issue where words fail. Mimicry too has that same 'beyond expectations' effect. It illustrates sensing in a perplexing way.

'Mother care' in plants consists solely of seeing to it that seeds are well equipped for their big journey into the unknown. It is the only voyage in any plant's life. That there is any care from 'home base', however weird in our eyes, is interesting enough, proving plants to be far from senseless.

Plants on the whole have three life-worthy connections, namely with the sun's processed light, their root system and, thirdly, their creation of flowers in most cases, so unlike the animals. It shows the enormous importance of plants. Plants lead a 'double life' so to speak, in the sense that they exist all by themselves without any 'shopping around' and they are at the same time 'food packages' in green for animals (or not so green as the case may be). A feel for it must therefore have been responsible for it all including the creation of the universe, and it became merely very refined in us, in our 'gifts'. It is altogether a mind over matter issue, like those who moved the huge stones of Stonehenge. A mastermind behind it all therefore cannot be ruled out. What we are seeing now, in the universe and life, is a huge magnification of it all or, in other words, a refined impression. Both issues have in common that they expanded in an explosive manner. A description of it as awe-inspiring is therefore justified.

Life represents a realm of which we don't know the boundaries. It concerns two kingdoms, neither being quite of this world – however much they may be expressed as 'worlds' instead. Earth is solely a pied-à-terre for life. Were Earth to die one day (hopefully of old age), it would be too ridiculous to assume that life would too. Earth is a realm with strict boundaries – so unlike life.

A bee orchid is mainly a matter of 'To be or not to be'! The creation of the plant, which included a very serious processing from light to chlorophyll, was clear evidence it was going to involve the entire world – Earth – with life in quantity and quality. The big question is, what triggered it all off to include part of sunlight, of all things, in the overall living process? Somehow it, a given part of sunlight, turned out to be a very important link. It is most extraordinary that something like sunlight, once processed, could be so powerful that it is capable of creating an entirely living world, which moreover consists of two entire different realms. How could a very simple life form possibly have incorporated sunlight in the living process? It was just too far-fetched and it must have been, therefore, a super-power causing it. Not only was the assimilation process a novelty by any standard, but it was also extremely important, resulting in a worldwide living expression. That processing was of vital importance to open the way for evolution.

The bee orchid, as an outstanding example, is clear evidence that extreme sensitivity is behind it all, which includes the actual origin and the most enigmatic creation, that of the plant world.

Plants are created synthetically to a great extent, absorbing light by sensing. A plant is mostly a matter of feeling its way underground and above ground.

Plants not only lead a 'double life', because partly they are a kingdom all by themselves, save for some assistance and food provided on the other hand by all animals (directly or indirectly), but also because they exist partly underground and partly above ground.

It is via the sun that plant life continues almost entirely synthetically.

Mimicry is a veil lifted, showing there is much more to it. The out of this world processing, from sunlight to chlorophyll, resulting in plant life, was clear evidence that a super-might was involved, resulting in life spreading eventually over the entire globe. It is sensitivity in the extreme which made plants create flowers. It concerns a 'feel' which goes much deeper than we can fathom. It is that kind of abstract sense which must have been present from the very beginning. Plants are the ones with the extras, namely the involvement of sunlight, their rooting system and their flowers. Plants flourished through their flowers, literally and figuratively. All three specialities of plant life materialised due to super-sensitivity. So much is realised from light to chlorophyll. It resulted in a dual living world.

That potential in sunlight is a far-fetched one to say the least, but one which we all depend on. It is a sobering though that without slimy algae we wouldn't have stood a chance. It is also a sobering thought that sense or 'a feel for it' – in the language of an artist – went as far as all that to make creation on Earth work. Drawing this a bit further, being far-fetched has a big share in it all concerning the universe and life, with sense as the driving force. Spontaneity does not fit in this concept. Sense shows to be in all this a spiritual super-might. It made mimicry for one thing. This proves that sense is behind it all. It is most amazing that plants managed to create flowers for the sole purpose of attracting members of the other living world to achieve fertilisation. In that case too there must be some kind of mind at work in their realm.

In other words, the saying 'it makes sense' applies to the plant world, however unlikely that may sound. It proves that both living worlds have one basis of origin. It makes flowers bloom. It also gives strongly the impression it was all pre-planned and the flower theme meant a huge increase in the plant world. This world was ready for plants – and for us, in the long run.

Plants are created by the mysterious assimilation process. How extremely delicate life on Earth is, shown in the huge variety of plants and animals, yet on the other hand both have been holding out over billions of years. Plants were the first evidence that life on Earth wasn't going to consist of very simple life forms. In plants, the project goes from abstract (sun rays) to matter (chlorophyll). It gives strongly the impression that the road from abstract to matter has at least a very great deal to do with it and realised both the universe and life. Plants somehow also sensed what's good for animals. This is why there are apples, pears, nuts and blackberries, among many others.

Life turning green in one section and almost from the onset indicates its make was far from entirely earthly. Light and life are in reality in very close contact with each other, especially in plants, since light plays a very large part in creating them. It turned out to be the other dimension vital to plants. Plants only make life even more of an 'unlikely story' – and if the one was to be a feasibility nonetheless like life on Earth being very small at the beginning; then anything is possible, even issues beyond normal probabilities like plants. Plants made that 'field' very 'outlandish'. Plants are involved in living on Earth in a very direct sense, rooted in earth as they are and yet, on the other hand, they couldn't be leading a more out of this world existence, being in intimate contact with sunlight.

It was for a very good reason that sunlight was chosen to have its share in creation. It was to be the most important participant in the life of plants. The question remains: what made a very simple life unit at the beginning of it all draw that conclusion? It turned out to have the strength to build a life force the size of this world. Apparently there was that enormous potential in sunlight and it led to, among other things, the massive biomass of tree trunks, but especially the creation of flowers. That quality of light was sufficiently powerful to create not only one but two living worlds. Sunlight is, therefore, valuable for its light and warmth passively and actively in plants' case. Light, in other words, is a three-dimensional power, supplying light, warmth and, above all, life. It has the effect that both places, Earth and Sun, are responsible for sustaining powerful life. Processed light plays a major role in plant life or, in other words, plant life is light-based to a very great extent. To use sunlight as plants do is so far-fetched that it must have been pre-planned from the very beginning. Plants realised life twice over by creating their own kingdom and indirectly that of the animals. Plants made it all happen – on Earth at least. In plants' case too, sense, as a driving force, was there all along and especially clearly expressed in the sensitive plant.

A telecommunication developed between the plant and sunlight. The plant, therefore, was an extraordinary response. Plants have been around so much longer than we have and, therefore, one wonders when we will truly 'blossom', leaving our 'teething trouble' behind.

Animals are known as creatures, regardless of their size or shape, however sophisticated they are in many cases. This happens to be the case from the earlier forms onwards – setting plants apart for starters, although they too are creatures (created images), strictly speaking. They are usually not considered as beings and it makes them therefore a race all of their own. It resulted in two very distinct groups, yet from the same stock originally. It is furthermore true to say that, without this big partition, life on Earth would never have materialised. It went to that extraordinary length by literally incorporating the sun's rays in the actual living process. That was a must to develop life throughout the whole world.

Plant life is living of another kind, just as effective but much more complicated, because of that extraordinary green biomass. It had to be a speciality since it came to a complete standstill in anchored plants. The plant has paper-thin leaves, while animals have bodies, which is just another example of the sharp difference between the two worlds. There is a good deal going on in plant life, and that makes it well worthwhile to look at them from a philosophical point of view. It is most extraordinary they ever came into existence at all. It just shows how important the sun is in it all, namely in the life-giving process concerning wildlife on Earth. Plants have their skyscrapers too, in trees, and, what's more, not from borrowed material. Many seeds give a remarkably close impression of being tiny spaceships, crammed ('manned') with data. It is also a plant's only journey into the unknown during its lifetime. It is here sense again that made it happen and, without it, plants would have been lost. This motive of sense is part of life and has been there from the beginning. Being rooted to the spot in plants' case gave rise to flowering above ground. There was no other alternative in the plant world. A plant exists between three stations, namely the rooting system, the flowering part, but, above all, the sun's rays. An animal is usually a body with a head, able to lift off the ground by means of legs. Plants on the other hand simply grew upright eventually, magnified in trees that way, and stand still all their lives – although this is far from the case metaphorically speaking! This abstract attraction is what leads ultimately to seed.

Our own development from caveman to, for example, civil engineer, is as phenomenal as that from algae to bee orchid. Both are mind over matter issues, but it turns out to be more puzzling when bearing in mind that plants lack brain power.

Plants may be an out of this world creation due to their use of sunlight in their living process, but they were to be the great breakthrough for all life on Earth. Plants too, once dead and after leaf mould, turn into abstract again. Furthermore, plants' involvement of sunlight in their existence shows the close relationship of abstract in the overall living pattern and the creation of flowers is another outstanding example. The latter was realised by an extremely high degree of sensitivity. Plants are a matter of holding tight in roots, getting airborne in seeds often and processing part of sunlight, all for their survival.

A flower and, indeed, chlorophyll are both in their own right more than merely a realisation via evolution. With that sort of power behind it, in plants' case mastering the sun's light, no wonder they blossomed with flowers!

We take the concept of a bee visiting a flower for granted and see nothing special in it. Plants, in fact, come closest to welcoming visitors from another world. Both chlorophyll and flower are created by being in very close touch with two entirely different worlds. Chlorophyll by having incorporated this world, the sun, and flower, with the animal world in mind. They are extraordinary issues by any standard, with the emphasis on *extra*. In other words, plants with their photosynthetic process have extended the 'field' of life beyond that of restricted Earth. Plants went into the depth of matters literally by means of rooting and figuratively by involving part of the sun's rays. Somehow there was that immense urge to go for sunlight.

Plants are attracted to sunlight by means of an abstract impulse and sunlight is equally abstract, and that results in biomass chlorophyll once the assimilation process is under way.

In our own lives too abstract plays a major role, as in abstract thoughts, which may lead to inventions. It shows the enormous importance of abstract in life on Earth, and animals at large wouldn't be here without plant life.

For example, an apple didn't materialise because of evolution in the first place, but by sensing, and so did mimicry. The latter is definitely not just a whim or luck, but business as usual. In mimicry something surfaces which wouldn't look right even in a fantasy story! It is simply too far-fetched to be taken seriously until life shows otherwise. Chlorophyll is where abstract (light) meets reality (matter) and equally mimicry is where 'wishful thinking' materialised.

Mimicry in plants' case is somehow 'seeing beyond the eye'. A cutting, stuck in the ground and turning into just another tree eventually, is the nearest example of an out of this world manifestation. It concerns a branching out beyond the reach of evolution. A plant is therefore extraordinary because of its chlorophyll, without which there simply wouldn't be a plant.

Plants read life quite differently or, in other words, their approach of dealing with survival is by means of 'going green', which is only a colour to us.

Any life, whether plant or animal, did not come about by evolution in the first place, but by a registering – mimicry expresses it emphatically and so do we. Plants are not just life in green, but emphasise the puzzling past. They are the enigmatic ones *par excellence*. Plants need both, Earth and Sun to perform. The former to penetrate into partially and the latter to absorb an abstract ingredient from. Both are necessary for life.

Plants move home in the pre-offspring stage as seeds – which is called the egg stage in animals. Plants' origin can be traced – just. Life's origin goes one step further.

Plants exploded into a living world, which includes sky-scraping trees, and came to enigmatic flowering all because of the creation of leaves.

Abstract sunlight turned out to be part of actual living known as life on Earth.

Plants are abstract because of light on the one hand yet firmly rooted in *terra firma* on the other hand. It is a contrast in the absolute sense.

Immobility in plants led to flowers consequently and that means there is a strong capability of sense in them as well, as is the case in mimicry, plant-wise. The latter is perplexingly expressed in a bee orchid. It is just baffling! In short: a fraction of the sun's light made it all happen on land.

Sense is creation in depth. It also proves that there is much more to life than just growth. It puts colour in it all.

Plants are the ones which seem to have come from nowhere. In other words, they give the impression of having been created out of thin air mainly. Plants do not have much to go by, and they have to be satisfied with a crop of leaves as the main issue in their existence, besides the flowering section of the upgraded blossoming kind, which almost consists of a world all by itself, known as wild flowers. Leaves come in an enormous variety of types, from the needles of evergreens to water lily leaf. A leaf is surface matter in green and it really ought to come under the heading of 'beyond expectations'. Imagination, inspiration and materialisation is a line followed up in mimicry, which, again, is a matter of contact made with abstract. Abstract, therefore, seems to be always close at hand. All this means that so much in life comes close to mind over matter. Indeed, the same seems to be the case in the universe at large, where little – if indeed anything – in the way of accidents on a huge scale seems to happen. Living green, as in plants' case, due to light, comes very close to mind over matter.

A plant is an absorbing system. It is a matter of absorbing part of sunlight, the goodness out of soil and a kind of absorbing (attracting) of creatures from the other world for the sake of their flower fertilisation-process.

It was precision work of the first order that resulted in selecting the sunlight to be processed, which had the effect of creating all land life on Earth. In that case there was a super-might behind it all – four-billion-or-so years ago.

It is all because of chlorophyll, basically, that there is us and the countryside. I lived when I was young in a surrounding of beautiful countryside, about half an hour from Amsterdam by train. It was called the Eng and there were big areas of heather and woods. I didn't realise then it was all basically due to chlorophyll that plants 'saw the light', metaphorically speaking. They made contact with abstract and turned it into matter via a chemical reaction. They had their own reading, as it were, of *veni, vidi, vici*. It proves that light happened to be a potential of the first order. In other words, plants are so out of the ordinary that it is most amazing they ever came into being. However, without them, life would have been merely wishful thinking.

Butterflies and moths manage to pick up one plant 'station' out of many, like heather for the emperor moth, nettle for the tortoiseshell, wild carrot for the swallowtail, and so on.

It is sense again in plants and animals when it comes to picking up aromas by insects and others. Hawk moths are even named after their foster plants, like the privet hawk moth. Plants are, for them, radio stations to be tuned into. They became used to it long before our time. Migration is also closely related to a tuning in.

Plants are an expression of 'reaching-out' *par excellence*. It is in their case a matter of reaching out by means of fragrances, reaching out for goodness in the ground they stand on and reaching out for animals in general with their fruits, nuts and many seeds among others. It is interesting to realise that bush growth everywhere, which is animals' homes and food supplies, owes its existence to the sun, since plants would be non-starters without photosynthesis.

Aromas, tastes, poisons, stings, juices of fruits, the 'meat' of nuts and even rewards like nectar are all attributes plants created for creatures they cannot even see. It proves again how everything depends on sense. Plants, so unlike animals, have to go through a process which concerns part of sunlight, for the sake of existing. All life on Earth depends on it. In short: plants are realised in a most peculiar way. The leaf came to life as a result of this 'link up' of sun and earth.

The creation of each individual on Earth owes its existence to the abstract in light. In other words, without light there wouldn't be plants – or us. Plants are the most extraordinary creations imaginable due to processing part of sunlight for their existence.

It is the 'ticking over' part (even in plants) that led to the bee orchid, among others. Like plants producing chlorophyll – the green mass – grey mass is the substance that made us the race we are. Grey mass even owes its existence to green mass since, without plants, we wouldn't ever have materialised. The link to the sun is especially clearly expressed in flora, the green world – at least, here on Earth. In other words, plants have it both ways, and Earth as well, as the sun provide a vital goodness for them. Plants are a joint creation of the earth and the sun.

Plants are responsible for all life on Earth, while they them-selves owe their existence to a mysterious life core at the very beginning of it all.

Most flowers are unparalleled beauties of an unknown design, which means we cannot read them entirely. There is not a shadow of a doubt that light has a great share in providing Earth's land masses with a cloak of green. It is in a sense more a godly display than the creation of us who, despite our sapient quality, merely came along in good time. Green itself is just another colour to us, like red or blue.

However, the might of green in the sun's rays underwent a processing accordingly. In plants at any rate, light and life come very close together. Indeed, it is where they turn into a bond. Plants, in a way, are a heavenly mixture in which it is difficult to say where light ends and life begins. In other words, light paradoxically keeps us in the dark to explain where it comes from. It has in common with life that it is there, as much as it is inexplicable deep down. Plants are profoundly light-involved and one part is processed into chlorophyll. Light, a special part of it, was going to be of major importance, to spread wildlife all over the world. In other words, a special section of light was the major factor to make it all happen.

There must be in a given part of sunlight an abstract strength that resulted in a worldwide spread of life. There is at any rate a creative power in part of sunlight. This power must be apparent in other places in the universe. In short: light means life and light is universal. Part of sunlight has a potentiality more important than anything else for life on Earth. Light, passively and actively, made it all come true. Life is an extension of the very beginning and part of it is an active participation in the creation of plants. Light in other words is a source of universal power, which life depends on. It might well have played its part in the actual creation at the very beginning of life on Earth. Light at any rate was present at the very beginning of the creation of the universe. It shows that light is a super-power of the first order. Sunlight also results in daylight, which almost no life can do without. Plants realised life upon Earth and they indirectly created us.

Plants are green proof that most of life is a matter of materialisation by realisation; the same realisation which leads us to say: 'I see'.

Mimicry is a development beyond earthly experiences, like a bee orchid, or a spider looking like an ant, and such cases provide room for more life-forms. It merely surfaces more clearly in mimicry. Plants prove sunlight has a quality additional to light and warmth in general; it merely puts another much more profound light on… light!

It is, therefore, light's power which plays a major role in life on Earth, passively and even actively to a degree. Light is as old as the universe itself. Plants and light are very close and even actively in contact, as far as photosynthesis is concerned – most extraordinary. Life is therefore partly light-created, and expressed primarily in plant life. Light made it all possible, quite apart from providing daylight. Light, in other words, realised plants to a very great extent and roots came second in this. Light also subsequently resulted in plants creating flowers, which is a very special extension in their life patterns. Light has more importance than merely supplying light.

We were more a matter of 'bound to happen' than from plant to flower – which means that plants deserve another hard look. We are a kingdom of a kind, like those of plants and animals, but we are lacking one king, and United Nations only sounds perfect on paper.

Plants made the impossible happen. They bridged the gap between abstract and biomass with their chlorophyll. They proved also that there was much more to life on Earth. They were the most unlikely creation, but they turned out to be of enormous importance for a worldwide spread of wildlife. In other words, they made life a reality for animals and, therefore, they have a most important reason to be here. It also made sense to grow beautiful wild flowers. Plants 'saw the light' without being provided with a pair of eyes. They proved sensitivity is tops – and even beats a pair of eyes in important issues. It happens to be the same feature enabling a blind person to say: 'I see' (metaphorically speaking)! Plants put a finer point on attraction. The flower materialised due to an imagination *par excellence* – and it is us who are baffled by it all.

It gives the impression of biomass (life-mass) and light-mass being stations not that remote from each other. Light must also have had a great deal to do with that first and most famous explosion, known as the big bang. At any rate, lightning and thunder are closely related. The lightning itself gets its energy from a mysterious build-up, which itself started in that contrasting 'stillness before the storm'.

Plants, at any rate, put another light on light, and it is a much more important feature than we know as light. Without that plant knowledge, there wouldn't have been any point about anything. There is a tell-tale finger pointing in plants at light being responsible for a great deal in creating – and not just plants. It is a universal attribute and as such has had a great share in creating the universe. In short: there is much more to light than comes from merely looking at light! It is an abstract extraordinary. Without this 'extra bit' everywhere, which surfaces so clearly in mimicry, there wouldn't have been two kingdoms – indeed, no life at all. Plants may not live by thin air alone either, but they certainly do get closest in that respect.

A rose talks – at least, one did to St Francis – but so does a stinging nettle and its 'language' is very plain indeed. It all stands out as quite baffling, bearing in mind that the starting point was slimy algae. Once light was absorbed by living plants, they were bound to develop via evolution, but that still didn't include flowers. These are creations within a creation (as with mimicry). A flower, again, in a green plant, is a matter of light functioning to a very great extent. It is 'alien' light here where it is an 'alien' bee later on for the flower. Both are absorbed in a way. Abstract apparently is always very close at hand indeed – in light or sensitivity. It could be said that bodies are the outcome of abstract influence in a three-dimensional way.

A rose, both lovely and poignant (in its thorns) as a shrub represents values which could only have materialised via a registering, which is in principle the same feature found in us, albeit in an enlarged manner. This registering or responding is found anywhere in the universe.

Life turned explosive by plants touching the abstract in sunlight. It is that green touch which is responsible for so much. It is a switching-on to a green wavelength, which could only have been realised by a super-power. It concerns extraordinary sensitivity and it is the latter first and foremost which created life on Earth.

The creation of plants was about as big an event as the first ever one. Without it, any creation at all might as well have been a non-starter. It would have been a matter of 'mission failed'; but it is now 'mission accomplished'.

Plants are a display of sensitivity, emphasised in their flowers.

Mimicry and attraction, the latter on which so much depends, are both issues where tuning in is of paramount importance.

It is quite clear that life shows great interest in anything of abstract matter, like sensitivity, attraction, sunlight (especially a given part of it), and even mimicry. There are no less than two living worlds in their entirety depending on the sun's rays for survival, with that of the animals being in an indirect sense.

The creation of plants was clear evidence that life on Earth was going to be a huge undertaking. It gives strongly the impression it was meant that way, seeing the enormous complexity of the processes involved. It was much too complicated just to have been an odd chance. Were we depending on the sun as much as plants are, we would have been entirely different as individuals. As it is, plants are almost part of us – yet they need no supplement from us; they are a kingdom all of their own.

Plants may not be able to talk, but all the same it is them which 'say it with flowers'. In short, a flower is a plant expressing itself, usually in a very colourful way. There is, of course, a much deeper undertone and a flower is a tangible expression of sensitivity or, in other words, a flower is an object which makes a good deal of sense. It is the same sense in action, which expresses itself in 'where there is a will, there is a way'.

We have distanced ourselves from the other two kingdoms to such an extent that they almost resemble alien worlds to us. Mention ought to be made here that neither of the two kingdoms has a king as a ruler. This distancing by us from other animals and plants originated when we left the woods for good – and in an upright manner (though it may still be a good place for a picnic!). We literally walked away from it. However, although there is now such a gap (and something like a motorway only emphasises that split) we are also much better equipped on the other hand, enabling us quite paradoxically to penetrate much deeper into their separate worlds – from afar as it were.

There is also another remoteness (between Sun and Earth), which led to so much closeness in chlorophyll. Leaves had an enormous part to play in the development and expansion of light. Plants are 'switched on' by the sun's light – or at least part of it – and in turn 'switched on' our 'light' via brain power eventually. Plants were a must, or life for us would never have been realised. No wonder plants anchored themselves in earth on Earth. They were that important!

It is all a matter of so much depending on so little, namely from light to chlorophyll up to a huge living world. Plants reached the giant stage themselves in trees and definitely also in a three-dimensional way as expressed in flowers.

It is very interesting to realise that a valley looks especially beautiful drenched in sunlight, the very light which had a huge share in actually creating it.

A flower is also a symbol of making the best of all potentials around, emphatically expressed that way in its beauty.

Making sense stands out in mimicry, making us sometimes laugh at its perplexity.

Evolution, with all due respect, is no more than a by-product and sense is the main issue. Sense is the drive behind evolution (both being abstract values). Sense also led to us, via chlorophyll in plants. In other words, there wouldn't have been evolution if there hadn't been sensing in the first place.

Sense made evolution make sense. It was also sense which made contact with the planet from the start, and there again it is sense that was attracted to light and led to processing resulting in chlorophyll. It was a huge undertaking at the beginning to create chlorophyll from light in photosynthesis, which was clear evidence that life was going to be a very serious activity on this planet. Plants turned out to be very explosive in their spread worldwide once light began to be processed.

Whatever the case, the creation of the plant was a most extraordinary effort, the more so since it happened close to the beginning of it all, when life, in so far as there was any, was still very primitive. In other words, the last thing that looked like an option was, of course, the use of part of sunlight – for the sake of living by means of a processing and consequently resulting in chlorophyll. It was nevertheless the ingredient necessary to create an entire kingdom. In short, it went from light to life. Almost certainly it is a matter of part of sunlight excluding the light (as in the case of plants) which has had at least its share in the first signs of life on Earth. In other words, a photosynthetic involvement might therefore have been present at the arrival of the first signs of life on Earth. The fact is we are all still strongly sunlight dependent.

Life itself has been a matter of *veni, vidi, vici*. It is also strongly expressed in plants that way, and seeds usually travel over a wide area. Plants which originally had so little to go by, immobile as they are, paradoxically made the most of travel – if not a downright contradiction in terms! This travel feature is concentrated in flowers. They draw bees, among others, towards them and send out seeds later, more than once in quite a violent way, in a catapulting performance.

A certain force must have created that very complicated species of plant. It was so out of the ordinary that it is most amazing it ever occurred. Once created, it showed a huge power and plants became an entire world, despite the fact that they literally can't move, emphasised by the fact that they are anchored to the spot by their rooting systems.

The urge to be photosynthetic, as in the case of plants, resulted in life on Earth. Plants process light to chlorophyll for survival.

However far-fetched, plants found a way to cope with the near-impossible. Plants' greatest creation is without doubt the well-developed flower. It is an expression of sensitivity in the extreme and beautifully illustrated to such an extent that they have our greatest admiration, with us being the most highly developed species on this planet. Apart from that, flowers are also most ingenious inventions. They manage to draw species out of the other world towards them by means of fragrances for fertilisation. It is the more bewildering in a way since plants making sense sounds totally out of this world. However, they are of much greater value than merely green growth. Plants enabled life on Earth to develop.

Immobility was a challenge (if ever there was one!) overcome by flower.

Without that most extraordinary assimilation process, from light to chlorophyll, there wouldn't have been life on Earth, save perhaps for some very primitive species. Light on the one hand, so very important to leaves, and 'feel' on the other hand, have contributed hugely in creating life on Earth. Due to plants, we are all part of the sun's light. Mimicry concerns two entirely different species, worlds apart, of which somehow one is tuned in to the other; this can only be described as phenomenal. Strictly speaking, there was no reason for mimicry and life would happily have evolved without it. Whatever the case, it is quite inventive and, therefore, mimicry must have been an issue with a much deeper undertone. It is the same hyper-sensitive tuning in which enables bats to find their prey. An expression like mimicry, being that of hyper-sensitivity and light, both being abstracts, played a huge role in realising the spread of life globally. Mimicry expresses a degree of sensitivity in the extreme, like plants tuned in to part of light. It was part of life from day one. Mimicry shows in its way what life is really made of. It absorbs more potentials and it shows to what extreme depths life went to express itself.

Mimicry, in other words, is an illustration of a three-dimensional and perplexing way of tuning in. It is a creation within a creation, like a flower. It is a happening of the totally unexpected – like a little moth, which looks like a bird dropping on a tree leaf (and which, when touched, drops to the ground.

Mimicry shows that life goes beyond imagination to express itself. It leaves us baffled. It is seeing double in a most incredible way. It shows the ultimate strength within life. It is all so logical, yet at the same time so far-fetched. It means that it went beyond limits. Mimicry made life more colourful, as flowers show already so clearly, and literally. Plants are where the big metamorphosis takes place, from abstract to matter.

Mimicry is another example of abstract turned into matter, like from light to chlorophyll. It is where words fail to express it entirely. It all boils down to the fact that where there is a will there is a way, and mimicry put a finer point to that. Plants' attraction is often realised by tuning in to bees in their flowers, which is so much the more extraordinary because insects live in a totally different world. Nevertheless, somehow plants are aware of them.

Plants had to be so much more inventive, being stationary, than animals in existing, and they created flowers, fruits, roots and seeds in packages of one kind or another. By involving the sun's light to the extent plants do, the whole issue is barely earthly any longer, fanatical earthlings (having root systems in the earth) as they may be at the other end. It is a living in extremes and the penetration power applied both ways.

Mimicry may be rather alone in its expressions but it is clear evidence of the sheer power of life all the same. Mimicry is a copying *par excellence* that could only be realised by a degree of sensitivity beyond our knowledge – and there again, mimicry is the odd one out as a totally unexpected way of expressing life. It is an exposure to the supernatural quality of life.

An out of this world urge realised the assimilation process, creating plants eventually. It is possibly the same urge that directed a life core to make contact with Earth. Life and light are closely related – life is light-involved to a great extent. Light has an important role in the life of plants. The sun, therefore, has its share in life on Earth, apart from providing daylight.

Plants had to be created, however immensely complicated, due to a process that is beyond our knowledge, being necessary to create life on Earth. It is this most intimate attraction towards light that, once processed, creates plants, as much as bees among others are attracted to their flowers. Plants utilise light in a mysterious way. Roots' role comes second in this. Light, in other words, is so much more than just light – at least, as plants have it. They prove it photosynthetically. In plants, the intangible becomes tangible, being the biggest metamorphosis of them all. By inventing chlorophyll, plants developed into a kingdom. We are, in a way, a motor driven by 'alien' fuel, that of plants.

To realise us, Homo sapiens, plants had to exist first. We are the paramount outcome of that 'mystery tour' over the aeons.

There are many ways in which awareness is displayed by plants – and a bee orchid is merely one outstanding example in that respect.

A spider catches insects. Everybody knows this, but does not consider that here we have a wingless creature upon whom it apparently dawned that not only are there winged insects, but that they could be caught by being netted – and they were.

A fraction of the sun's light was to be the major share in the overall creation of life on Earth.

It is due to the assimilation process that the boundaries of the plant kingdom lie beyond earthly dimensions.

Plants are quite organic in structure, but they are however in very close contact with inorganic material like soil and light. They 'feed' on both.

Plants, despite being 'nobodies', without the most essential shape to animals, are nevertheless not 'nobodies' at all; they are merely proof that life could also be realised without a body, heart or, indeed, even a head. Plants indicate that there is more to living than breathing and a heartbeat. They also show it can be done without a body, thanks to puzzling chlorophyll.

Plants have been homing in on Earth, literally, by means of

roots – yet plants are the exposed ones day and night, quite paradoxically. Plants go for the sun's light in a much more profound way than we do, despite our holidays to sunny resorts.

Plants may be nobodies in a literal sense, yet it is they nonetheless which blossom quite realistically in their flowers.

Plants pick up the sun's rays and respond as a receiving station.

Plants are a working unit and they would be categorised as a chemical plant in our industry, producing photosynthetically chlorophyll, of which all other departments on the premises depend, like trunk and roots, but above all the flower.

Plants blossom in their flowers and show quite clearly the supernatural strength of nature.

Plants 'weather it' day and night, outdoors in their case means all the way. Plants have no legs to stand on, yet it is them who 'stand their ground'. It is also them, as green mass, showing that grey mass does not necessarily have the monopoly – or a bee orchid would never have materialised.

Plants flower often in a magnificent way – yet have no eyes to admire this themselves.

Life is centred all around leaves, which just couldn't be flatter usually, or, in other words, plants realised life upon Earth. They were to be that important.

Plants are realised from a totally unexpected origin, namely out of sunlight. However, it led to a worldwide spread of flora and fauna; it even came to unexpected flowering that way and that is 'beyond expectations' twice over! Plants go to extremes much more than animals. They absorb goodness out of the soil they stand on, they face up to extreme weather on the spot, but, above all, they absorb sunlight for survival.

Plants are the place where the inorganic meets the organic, as demonstrated by light to plants. Workers like bees are introduced by means of attraction. It sounds so obvious, being non-travellers themselves – but how ingenious! Plants 'eyed' bees, blind as they are. Attraction has been glamorised in mimicry, which is an adaptation *in optima forma*. Deciduous trees lose their leaves well before winter – and turn into skeletons *par excellence*.

Plants didn't just go to 'the bottom of it all', but went deeper and also underground.

Furthermore, plants provided 'the good earth' on Earth. Plants are as solid as can be, as giant trees, as well as being the closest to abstract in creating chlorophyll from light. Plants, in other words, have a way with life – on the spot!

The plant was created in a most extraordinary manner. It developed – and still is developing – in an out of this world fashion, namely by absorbing part of sunlight, which is processed and turns into chlorophyll, as the main part of a plant's life. It makes plants more outstanding still than animals.

Plants, lacking the mobility so familiar to animals, had to be therefore so much the more inventive to survive. It resulted eventually in the creation of flowers, which resulted in involving the other – to them, alien – world, namely that of bees. It comprises a huge paradox by forcing plants to find ways to run their 'business' smoothly all the same, despite the great handicap of being held back by roots. Plants got to know light from a different angle. They 'channel' a great deal. It shows in the stalk of a leaf or flower, or the pipelines of roots, the outsize 'stalk' known as a trunk in trees and in an abstract style in channelling part of sunlight. Plants have their restrictions. All are stationary and green (save for some exceptions) but, despite that, the variety on the theme of plant is more than sufficient to fill a kingdom. Plants, taken overall, possess that 'beyond expectations' impression. They express it in their chlorophyll, in their hibernating known as dormancy, in their flowers, which are after all extraordinary extensions, by life within earth in their root systems and by being homeless in the absolute sense, exposed as they are day and night.

Mimicry, in its variety, is tangible proof of supernatural sensitivity. So have flowers and also their creation been realised consequently.

Plants prove that sensitivity has been there all along and isn't merely part of the animals at large, glamorised in us.

Plants are the green fuel for animals' engines, including ours! 'Being in touch' went up to phenomenal strength in mimicry. The concept of 'close touch' stands out in plants quite clearly in a three-fold manner: in creating chlorophyll photosynthetically; in attracting a third party in many cases to achieve fertilisation; and last, but not least, the element of mimicry.

Plants were the Capability Greens, which enabled animals to reach the status of kingdom, and it is because of them that a Capability Brown would emerge one day! Plants are self-contained *in optima forma* (save for some assistance in the fertilisation period), and they do all the work 'at home'.

Plants live in extremes. They go into the ground they live on with their rooting system, absorbing goodness from the soil, and they absorb, as the other ultimate, part of sunlight which, once processed, turns into chlorophyll. Plants' drawback is being confined to one spot all their lives, which forced them to turn into 'exhibitionists' or 'extroverts' as shown in their flowering.

Plants govern a living world, including the mobile one. They are wildlife's green potential.

Plants are as much earthed in Earth as they are utilising part of light. It is this contrast in the extreme that all life on Earth depends on. Plants are intriguing and it makes them seem odd, being 'fed' by part of sunlight. However, if they hadn't done so, we wouldn't have existed.

Plants too have their *plan de campaign*, even literally. It resulted among other issues in the creation of flowers. Plants had 'an eye for it all', blind as they may be in the strictest sense. Their ultra-sensitivity made it all possible.

Plants prove the point beyond doubt that 'I see' doesn't necessarily have to translate as 'eyes see'.

Plants deserve the limelight by highlighting light. Plants are food, directly, as green (grass) or indirectly as seed (corn), fruit (apple) or root (beet) and, furthermore, a sweet shop for many in the case of numerous flowers. Plants show a way of existing in its most elementary form, unlike animals, where there is a buffering effect relying on plants. Plants, in other words, get closest despite starting from nothing. Plants are greatly involved in abstract. Chlorophyll owes its life to abstract. Plants attract bees among others and seeds usually leave the premises that way, namely by abstract air. Plants are light-minded above all, as merely another way of expressing life. Light and life are very close and even meet each other in plants as an extraordinary amalgamation. At any rate, abstract shows itself to be in great demand in it all as a Very Important Partner. Plants are immobile, which has as a counter-balance their flowering quality.

The transformation of light to chlorophyll is a process of huge importance. Chlorophyll is the predominant feature in it all. It was all a matter of so small a starting point (a fraction of sunlight), resulting in so great a worldwide spread. Chlorophyll is a basic substance that all life depends on.

Plants have more than normal interest in light and soil compared to animals. Plants linked up light and life or, in other words, plants made light a shareholder in actual life. Plants could only have been realised by sensing in the first place, and flowers merely stress the point.

Chlorophyll is an enigma as much as it is a creation, which all life depends on directly or indirectly. It therefore made it all come true, save for some very primitive life. Chlorophyll is also nature's greatest spontaneous creation.

In one sweep, it went from light to plant to animal. It is abstract light that brings plants into this world. It is amazing to say the least that life focused on processing part of sunlight at the beginning of its existence. The more so in view of the fact that the assimilation process is a highly complex issue. A feature like mimicry proves that intense sensitivity is present throughout life and such a quality must have spurred life on to make contact with part of sunlight. Flowers blossom and they again prove in a beautiful, colourful way that it is all a matter of sense in the superlative. Plants started off all by themselves and 'outside staff' like bees came much later. The question is, what made plants be created in the first place? Animals are to a very great extent merely a following-up of data given. Plants' extreme intimacy with light has naturally a great deal to do with plants' startling expressions in so many ways, like their flower creation. Plants 'think' of existence with light in mind – and that's a heavenly thought! Plants show also that Earth is merely a station, not a terminal. They clearly put a finer point on it all, last but not least, through chlorophyll.

Plants may have no vision but it was them who got in touch with light in a very intimate way and it was them who got in touch with bees, among others, to satisfy their requirements. It was also them, with no eyes for beauty, who created the most beautiful appearance conceivable, as seen in many a flower. It raises them up above ordinary earthlings – even us in that respect.

Light, incorporated within the living system through plants, was bound to have the most startling effects. It turns up time and again in the overall existing program. The biggest display of them all concerns us ourselves and we are sustained by light, indirectly through plants. Plants sustain themselves on sunlight and the richness of the earth. Both come in plentiful supply. They are respectively abstract and solid, but it is the former (light) being the drive behind it all. Flowers come all the way from light-fed chlorophyll. In other words, light therefore has been explored to the full in plants and it shows best in colourful flowers.

Due to light, plants came into their own and due to plants, animals developed into a kingdom too. Light is responsible for so much in a positive sense and, due to this abstract matter, quite indirectly, grey matter was born. The latter was to be the core for a great deal of ingenuity. Plants are where the sheer impossible takes place, namely the process of photosynthesis. Plants are most intimately in touch with light and, at the same time, based within terra firma. Plants are a sublime example of being used to living on this planet. They adapted in the extreme to a completely outdoor life.

In the case of animals, there is a 'middleman', namely plants. Plants know yet another extreme and where some may not grow much at all over years, others are like the 'mile a minute' (the famous Russian vine). The latter is an extreme example, yet plants make a lot of growth yearly. It is this hyper-sensitivity, also clearly expressed in plants, which created them, due to the involvement of part of sunlight and it is yet again clearly expressed later on in their perplexing development of mimicry, as shown in the bee orchid. In short: it goes beyond us!

This ultra-sensitivity led to life meeting light, resulting in that famous process. The long stalk of many a flower is, in itself, a matter of caring by the plant. The issue is stressed in a ground-hugging plant in its familiar rosette shape, like in the case of a dandelion. Animals' 'fuel' is green (plants) and plants themselves are 'fuelled' by light in the first place. It means that light to them is almighty. The planet comes second in this – not unlike a 'second helping'. Light and plants are, in other words, most intimate. Light, however abstract, is their 'building block' *in optima forma.*

Plants represent a bridge in green as shown in many of their expressions, like between the two bodies, Sun and Earth, joining the two living worlds together through bees and others, they stand between abstract and matter in their photosynthetic performance. They link up above and under by means of their root system. Furthermore, the gap is often bridged by means of a lengthy and very slender stalk. Last but not least, plants' seeds realise the stretch between A and B, which means spreading by air usually. Even the plant itself has, in the strictest sense, come about by 'outside help' from light, in stark contrast with the animals which are by nature an 'inside job'.

Plants turned Earth into a viable environment for life. They did so with nothing to go by but light, processed photosynthetically, and air to breath, enabling them that way to change weathered rock into rich soil as they went along. The latter could be considered an extraordinary metamorphosis, which came into being through a root system.

It was to be that quality of strength that made plants feed a whole world and also enabled them to blossom with beautiful and heavenly-smelling flowers often – all because of the most intimate involvement of light, indirectly. It is this connection with the other world (the sun) which enabled plants to create flowers in their full splendour. It is plants themselves that open the gate to a field without an earthly horizon, however earthly they may seem to be in their roots. Life is where the abstract meets matter, and plants merely stress this point.

Sunlight has a three-dimensional effect in plants' case. It serves growth, it provides warmth, but, above all, part of it is processed for the sake of living. Many flowers are as beautiful as they are baffling. The contrast between a pansy's plant and its huge, magnificent flower makes one wonder how the two ever met. Flowers are almost no longer plants. They are often at the end of a very long and slender stalk, which merely stresses the distance.

There are many most striking instances of awareness apparent in both flora and fauna. Plants for one 'say' it in their poignant stings from nettles among others, but above all in their splendidly colourful flowers – and then, as another example, there is mimicry too in their kingdom.

Plants' enormous importance is expressed in great depth in paper-thin leaves. All life on Earth depends on this green panel structure. It also shows the enormous influence that solar power has on the living world. We are all involved in this light-might, with the leaf as its mediator. In other words, light itself is brought into the limelight through plants. Plants are the green power behind it all, not unlike the brain in our inventions. The leaf is the green 'chip' crammed with data and it led to the ultimate botanical creation: the flower. Even a tree trunk, however massive and solid it may be, owes its existence to the green canopy above in the first place and, secondly, to the root system.

It is because of their very intimate contact with light that plants render an out of this world impression. It remains a mystery as to how plant and light ever came together on Earth. Plants' life is a light creation. In plant life's case, light, or at least part of it, penetrates within. In other words, plants prove there is more to light than just light. Light rules the living world. Light is a super-might out of this world, and plants prove it and elsewhere.

Plants summoned bees to their premises by means of 'sign language', all the way from the other world, no mean feat by any standard. They did so by means of their flowers, which are equipped with several attributes in most cases, meant to draw the attention of members of the animal kingdom. The turnout, accordingly, is overwhelming. It is proof that it works most successfully. Somehow plants are aware of bees. Once there was life on Earth, plants came, due to light. Light is a source of vital importance, besides providing daylight. In short: fauna partly 'absorbs' plants which, in turn, partly 'absorb' light for their existence. Plants are attracted to light on the one hand and they themselves attract bees on the other hand.

It went from light to chlorophyll, and without this processing there wouldn't have been any life on Earth, except perhaps for some microbes.

To create chlorophyll involves the sun's light. The highly

complex creation of chlorophyll was a necessity, almost from the beginning, for life on Earth.

Mimicry is proof of awareness within life, but then so is a flower or, indeed, thorns on a rose bush or bramble.

There was a starting point in plants' case before evolution set in, namely the most remarkable process of assimilation, however that came into existence. The eventual outcome was a huge magnitude of life worldwide. Somehow it must have registered in the case of a very primitive but also highly sensitive life form, that getting in touch with part of sunlight, via an assimilation process, would lead to a huge living world. It is as if there was a master-mind behind it all, because life was only present in a very primitive way when chlorophyll was created, the more so since so much was going to depend on this extraordinary process. In short: plants came into being at the right moment, when so much was going to depend on this out of this world development. The plant was a complicated creation all by itself without the involvement of evolution. It is the more amazing it ever happened, since there was so little to go by at the beginning. It sounds too far-fetched ever to have been realised, especially at the start of it all. Whatever landed here on Earth must therefore have been something very special. In other words, it wasn't just an ordinary life germ that made contact with Earth. The plant was the necessary link for a worldwide spread of wildlife on Earth.

Plants, despite their very humble beginning, were a most extraordinary enigmatic creation, and were designed to be a huge green might. A leaf owes its existence to part of the sun's light. There was an urge to pick up this light to be processed for the sake of life, thus creating what we know as a plant.

It is most remarkable that it ever came to a joint effort between, of all things, sunlight and life, which led to a processing issue and created the plant. The word 'plant' itself means spread and it has all been achieved on the spot, due to a rooting system.

Attraction in flora fashion is of a higher plane still than in fauna and is meant for members of this kingdom, like bees. Dramatically put, a flower, in most cases, is the place where two super-powers meet, those of flora and fauna. This 'bee/flower line' concerns a sensitivity far beyond any human quality. If we could share it with them, we would actually 'feel' another world far away. A flower is, therefore, a shape with a much deeper undertone, impossible to be fathomed by us entirely. A flower is a developed expression of common sense.

A beautiful flower is not just a show-off: there is more to it. Flowers are 'beyond expectations' in the true sense. Flowers have a 'say' of which ours is merely an echo at best when we 'say it with flowers'. Plants turned into a world concern and all basically due to leaves. Plants don't grow merely towards the light; they are in fact light-bound at least as much as they are earthbound. Either way, it concerns a penetration into matter which is out of animals' reach. Plants are, therefore, themselves absorbing agents, expressed in their leaves and roots. This absorbing twice over of foreign matter, so characteristic in plants, gives rise to the growth from leaf to trunk to tree and from bud to flower.

Light may be there in a never-ending quantum, but the question remained nonetheless as to how to obtain part of this abstract quality as a potentiality – until a way was found, which led to an entire kingdom in green.

Plant is where abstract meets the tangible halfway, due to life, the big generator. Plants do deserve to be investigated by us. After all, we owe our existence to them, albeit in a very indirect sense.

Plants came at the right time, when the living world was at its beginning. What makes it so special is that it was a very unique feature. Plants had to materialise in order to realise two worldwide living units. The creation of plants is as much a big mystery as the actual contact life made with this planet. The extraordinary thing is that plants came along at the right moment. Plants were a breakthrough. Issues like the creation of chlorophyll, that of flowers or, indeed, thorns on a rose bush, mimicry and plants' rooting system, among others, all has to do with a hyper-sensitivity, which is beyond us.

Flowers, plants' own creation, made life flourish on Earth. Plants make Earth's bounds much more loosely woven. In other words, they have an alien streak in them. We can behold flowers' beauty, as is the case so often, but we are unable to explain wholly satisfactorily the beauty of it all in trying to explain how a relatively simple green plant could ever have created so much.

The sting of a nettle is already fascinating as an ingenious supplement to a leaf, but this, the above mentioned, goes so much further afield still.

A bee orchid is where a beautiful flower has been changed so that not only does it look exactly like a bumblebee, but it even smells like one. Males of this species come flying in from afar and even try to have sex with such a flower. This is so much the more remarkable since such sincere intimacy is not at all like plants.

The bond between Sun and Earth, established through plants, was bound to cause a stir, to put it mildly. It resulted in a big show in which beautiful flowers put a finer point to it all. The plant was the beginning of a dream come true. The big question is: what made early life possible for plants? It is the more important since they were essential. Animals gather data all along during evolution. Plants, on the other hand, were purely, basically created at the very beginning, due to the assimilation process. It was a matter of ingenuity in the extreme. It is most amazing that it actually happened! It was a matter of so much from so little to start off with, sufficiently to develop an entire living world all of its own. The animal kingdom concerns a huge variety of all sorts, while that of flora is a huge variety on one theme. It is because the sun is at the right spot that plants came into existence. They owe their lives to the sun, may it be in an indirect sense, and Earth came second.

The only remaining link between plant and flower is usually a delicate stretch we call a stalk, and even this is not very helpful because the two have just about nothing in common. The remoteness is much more pronounced as a rule than between body and head in the animal world, where the stretch separating them is usually a short neck, though a giraffe is a notable exception. Plants are as natural and organic as they are supernatural and linked up with inorganic light, which is photosynthetically dealt with within their living system.

That there was a force within light other than its shining ability perplexes us, realising this abstract matter processed would be transformed into tangible green biomass. We owe so much to the sun – indeed, even our existence, indirectly.

Plants too managed to hold their own over millions of years

despite the precision work required. Plants, in other words, are primarily an out of this world creation due to the intense involvement of sunlight, and they are secondarily earthbound. That makes plants worth a second hard look! Their variety is enormous and at least as great in its way as that of the animals, yet they are all chlorophyll-based and that makes them extraordinary. Once the chlorophyll was established, evolution played its part, so unlike the animals who depend entirely on evolution, gathering data by moving about. It makes it the more amazing that immobile plants found, paradoxically, their way. Their variety goes from grass, and even smaller examples, to sky-high trees. Leaves are usually paper-thin, but the might within is enormous and, among many other issues, result in plants' creation of flower. Sophisticated plants have managed to obtain 'home help' all the way from the other world in the form of bees and other insects.

Light is a subjective power in animal life, while in plant life a part of it is also an objective force of immense importance. In plants' case, life is highlighted from a totally different angle. It turned out to be most successful, however far-fetched. Life, in other words, needed light in a much more intimate way. An important bit was incorporated for that purpose. How it was realised that mighty strength was to be obtained from light to build a huge kingdom is anybody's guess. It is most amazing that a highly potential life unit landed here on Earth, which was a chance less than one in a billion – or was it all designed that way?

Plants 'made room', figuratively speaking, for the establishment of a totally different living world, that of animals. It originates from awareness again and extreme sensitivity that made plants flower. The latter is often well away from the actual plant by means of a long stalk, to increase its value dramatically. This is again structured due to strong awareness from the plant and it is a pipeline of vital importance. These abstract values of awareness and sensitivity must have been apparent from the very beginning of life on Earth. Many flowers are very striking (striking by being beautiful) and harbour goodness for their 'home help' like bees – all the way from the world of fauna. This, and much more besides, is what makes wildlife so interesting.

What makes it more interesting still is that wildlife is, in fact, a security *par excellence*, the reason why it is still around after all this time. Plants picked up data on the spot. It remains a mystery how some turned into gigantic trees. To plants, light is life-giving in a direct sense. A flower stalk is, in a way, good thinking material-ised. It serves firstly as a flagpole, secondly as a pipeline channelling soluble material to the flower and, furthermore, enables more efficient seed disposal later on. From seed to plant is the perfect *perpetuum mobile*. Life's outline in its myriad shapes and sizes is light-centred through plants. Life itself is focused on light. Plants are very sensitive to light. This sensitivity is as baffling as the creation of the bee orchid.

If there is so much 'life' in life, it could well be common practice in the universe; the more likely so since light, which is everywhere, appears to be so important in actual creation itself. To go for the sunlight was as exceptional as the actual life unit with its potentials making contact with Earth. It makes the animal world with its development by evolution seem straightforward in comparison. Plants made the living world what it is, including us, however indirectly. It comes down to the fact that an abstract quality was a necessity for realising life on Earth. A similar state of affairs might have had its effect at the beginning of it all. In short, life and abstract are closely related. It also makes attraction work, which leads to new life, including us and even many species of plants. Plants made their mark. Their existence was to be all-important.

Plants are the green might, so unlike the animals, which come in a variety of colours. Plants' origin is baffling to say the least. A massive tree trunk with a huge canopy above is a moving pillar in strong winds, which is beyond us! The plant was expressed due to an instant event, namely making contact with light – an event which all life was to depend on. It so happens that sunlight, once processed, realised viability. To negotiate full exposure as in the case of trees in wintertime becoming virtual skeletons is an extreme necessity to survive.

Plant life, in short, is focused on light energy first and fore-most. 'Born free', therefore, applies to plants especially, however much their roots might try to prove the contrary. In other words, plants on this world – even into this world with their root systems

– are in fact 'out of this world' through their most intimate contact with light.

Plants are immobile except for their seeds, which are often like small spacecraft, so unlike birds which go by air while their seeds (eggs) remain stationary.

Light, as an abstract, was partially to collaborate in realising life on Earth. In other words, life is light-dependent subjectively and even objectively in plants' case. Light is, therefore, as important as the original life unit to realise flora and fauna. Light is universal and may be so in life. If light is that important, it might well have made its hallmark elsewhere – or even in more places with a similar combination of two universal bodies. Light is as big a mystery as life itself. Both make a twosome at least as far as plants are concerned, apart from providing daylight. It is this joint effort of light and life that realised the existence of flora and fauna. Life in its original form, without light in an objective sense, would have led to nothing but possibly very primitive life forms. Light, or at least part of it, realised plants. It is far-fetched but true all the same that the sun is responsible for plant life. Light is a source beyond measure – like life. The life unit arrived at the right spot, known as Earth, with light's incorporation. Light, in short, is of vital importance. Light is everywhere in the universe and an assimilation process could exist somewhere else.

Light may be just an abstract but it is most amazing what it led to. Part of it is plant life on Earth. Plants' origin is realised by three abstracts, namely awareness, extreme sensitivity and, above all, sunlight. Plants absorb the supernatural in chlorophyll. The green line was to be the actual lifeline in the entire fabric known as the living world. Plants were the 'green light', the 'go ahead' for the traffic of life as a development on a world scale. It is as much far-fetched that the plant world came actually in its own as it is a most serious necessity. The process of photosynthesis is perplexing to say the least. Like mimicry, it concerns a sensitivity way beyond us. Something must have urged plants on to absorb it.

A tree shows itself to be an explosion of life in a silent manner, with its trunk as a column carrying the enormous mass above as a mushrooming cloud. To carry such a load, the trunk had to be massive as well as solid. This is a show of strength we have not been able to copy yet. A considerable amount of swaying is allowed during strong winds. It is the more outstanding still, bearing in mind that paper-thin and tiny leaves are responsible for it!

Plants combine light and the goodness of soil in their existence, which sounds like a fantasy to animals, the more so since light is abstract. Cutting a plant's flower means picking something which has its basis deeply rooted in abstract light. It means that any flower is, in principle, a sunflower. A colourful flower is a symbol of the paramount achievement of the green world's power. A leaf represents a show of strength in a very green fashion, paper-thin though it may be and quite small as a rule. It has no equal in life on Earth and its green form concerns no less than two kingdoms. It is perplexing that light itself was going to be involved in the big (living) plan. Light is plants' main income, figuratively speaking. The green part of light's primary function is to be involved in a processing which leads to plant life. Plants prove that there is a force within light to which life on Earth owes its existence – and light is everywhere in the universe! Plants originate from abstract light. Plants basically established life on Earth.

Life was meaningful from the start since a highly complicated unit had to be created, namely that of plants, to ascertain life on Earth in a worldwide fashion. The only way to establish life on Earth was by involving the sun in the process. Plants created a world within a world, namely that of flowers and, furthermore, indirectly provided room for another world, that of fauna. It is a sobering thought that all life owes its existence to chlorophyll, as expressed in leaves, which themselves are usually devoid of depth. From chlorophyll to leaf mould is but a step. It kept plants going through the ages, though. Plants show they can make do without their green cover in most cases over a long winter period. In the case of deciduous trees, only a skeleton is sufficient to see them through, until it is business as usual again.

Leaves are the go-between from light to flower. It is this annual dropping of leaves in autumn which comes so close to feigning death, as life-skeletons!

It is light which steals the limelight in plants' case and roots are merely their subordinates. Life on Earth, therefore, depends on the assimilation process of plants. Mimicry too shows in its way an abstract which materialised or an issue of wishful thinking that came true as an expression of mind over matter. Plants were an ingenuity in creation and animals followed suit by gathering data as they evolved. Plants realised life in two worlds. In plants, the out of this world image, namely abstract light, became part of the living world. Plants were the first to prove that whatever made contact with Earth was not just a little, unimportant project, since the outcome of flora, for one, showed itself to be a mighty power. Plants, therefore, are as enigmatic in their creation as the actual life core itself. Light in this, as a brilliant show of strength, has of course a great deal to do with plants' extraordinary performances, which reached their climax in colourful blossoming.

A flower is light-fed and leaf-fed. A beautiful one is where colour turned objective in a most awe-inspiring way often and is greatly admired consequently by us, the most highly developed species. A plant, as represented by a leaf, is a plain plan in which flimsiness has an enormous amount of depth. Leaves are plants' green panels and the forerunners of flowers. The green line, however far-fetched, was apparently a very worthy quest because it greened this planet to a very great extent and, besides, so much more depends on it. It gave rise to a combined effort known as life on Earth. Leaves are responsible for all life on Earth, directly or indirectly, and bees' (among others) activity is a mere part of it. Leaves represent a vast force sourced from light. They created, therefore, a super show known as life on Earth. A plant is, in other words, a Capability Green with an extraterrestrial tendency. It proves that life could only have come into its own here on Earth if that mysterious extraterrestrial part was included, linking us up with the entire universe. Life, in other words, is not entirely 'home-made'. Plants realised life on Earth in a dual greatness. They blossom in their flowers in the way we express a fast-growing concern. A colourful flower represents much more than merely another creation. They live, in fact, in an entirely different world and a long stalk enhances that effect of being set apart from the rest of the green world. It is all a matter of evolving.

Plants absorb light on the one hand for their livelihood and absorb soluble ingredients from underground on the other hand, whereas bees and the like, via flowers, turn it all into a life-triangle. Plants, in short, are a complete absorbing station.

It gives the impression that it could well be the case that life cores were actually made here, the sun and earth creating an excellent environment for such a happening, instead of crash-landing on this planet from somewhere far away in the form of meteors.

It is due to plants that we are here. They themselves came into existence as a result of a most complicated process which is an out of this world concept. It is the more outstanding because that's what has been taking place from close to the beginning of life on Earth, millions of years ago.

A bee orchid, as a very striking example of life's super-might, proves there is a kind of sensitivity within life on Earth which goes beyond our understanding. In plants, light meets life. Proof of awareness expressed in plants is a colourful flower, a stalk, fragrances meant for the other living world, a root system, the stings on a nettle and increasing acid in tree leaves if there is too much munching by caterpillars, just to mention a few points accordingly – not bad for green plants!

Plants are a green response and they are so close to abstract that they could find a foothold anywhere roughly like Earth, it seems, and animals would follow suit.

To fill Earth with life was a huge undertaking. A life form had therefore to be created, in which case a processing was to be involved from the beginning onwards and beyond earthly boundaries. The creation of plants was a response. The result is us – fauna at large are here also. It turned out to be a mission accomplished.

Deciduous trees have their own way of coping with winter, and simply mimic death in the form of skeletons. They thus see that period through and live to tell the tale.

Plants help us enormously in the decoding of actual life on this planet because there is so much green data available. Plants are a green providence, besides being a green realm. It is chlorophyll that opened the sluice gates to a flood of potentials. It meant, among other things, that plants themselves could reach the immensely important, colourful flower stage. Plants are so interesting as they were an absolute necessity to create life on Earth in the quality and quantity we see today.

Paper-thin leaves represent the blueprint (or 'greenprint') for life. They gave rise to the creation of a 'green house', being the plant kingdom. Chlorophyll in that world had a most original birth on Earth. Plants are the most astonishing creation of them all by involving part of sunlight in their lives, particularly when we consider that they came into being at the beginning of life on Earth. Chlorophyll makes it very plain that it is an issue too big to be grasped by us in its entirety, however tangible its outcome in leaves. Leaves, which are of vital importance to all life, are apparently not home-made, at least not entirely, and plants' rooting capacity comes definitely second in this. Life in general is centred around chlorophyll. In short: it went from light to plants to sight – and this seeing in itself has its most profound extension on insight as might! Both light and sight are abstracts, two super-powers in their own right. It all comes down to the fact that the mobile ones (animals) rely on the stationary ones (plants). Plants are the cause and animals the effect, which consequently created life on Earth. Earth is a mere station in the universe, with the result that we are all universal rather than being plain Earthlings, and plants emphasise this point clearly by being partly sunlight-made.

It is also true to say that sheer super-sensitivity, as emphasised in mimicry and flowers, was realised in plants.

From plant to flowery fragrances concerns a communicating with the other world, which in itself is evidence of green awareness.

Life on Earth is a plant-worthy undertaking and life is, therefore, leaf-based and leaves themselves are light-based. There was a response, or so it seems, at the beginning and that led to a creation due to sunlight, which resulted in plant life. It had to take place first and foremost to achieve a dual worldwide spread of life on Earth. It is all due to a hyper-sensitivity, as mimicry still shows us today, and flowers.

Plants, to put it bluntly, are a processed part of light creation. In other words, plants and light are not only close, they are a twosome. Light in plants' case means from abstract mass to biomass. It is this extraterrestrial approach that is so strongly expressed in plants. Light's own 'home' is outer space and it is this abstract which is of greater importance to plants than the actual pied-à-terre underground. It therefore puts into question the notion of plants being earthlings.

Another reading of plant life is that they are usually almost see-through in their paper-thin leaves. Even up to tree size, it is in plants in general, however flimsy their leaves, that an entire living world has been realised, being enigmatically chlorophyll-based.

Contact made by a life unit would have been pretty useless, without being followed by plants' extraordinary creation. Plants, therefore, are not only a huge and very interesting kingdom, they are an absolute necessity. Due to plants, things went from light on Earth to life on Earth. There is an amalgamation apparent in the plant and animal worlds which involves outer space, expressed in light. Life on Earth, in other words, is put on a broad basis; Earth alone could not provide.

The origin of plants was a sensational creation. We may not be green but we depend on green biomass for life. Plants are life with a green overtone as a necessity and, without it, I wouldn't be writing this book – moreover, we wouldn't be here! Plants being here is as baffling as it is true.

Light's most intimate interference in the overall design of things provides life with a universal hallmark, rather than merely an earthly one. In short, existence is a matter of the almighty being light-mighty.

The very best in green growth was apparently not good enough and a colourful display had to be introduced as the next in line, though it meant overstepping borderlines and flowers indeed, in their splendour, are 'way out'! It is sufficiently remote that it is almost incomprehensible; it is still plants' work, as often expressed by the phrase 'all in a day's work', since many a flower lasts only such a fleeting time. It is all a matter of being uncanny in green. A colourful flower and plain moss concern a much greater remoteness in expression than woodlice and even us.

That famous process of assimilation resulted in a worldwide spread of plants and animals. Being photosynthetically made means more than the very best Earth had to offer, which is goodness out of the soil for plants' roots. Plants are capable of attracting fauna to many of their flowers, like bees. It has the effect of the interplanetary happening here on Earth on a miniature scale.

To plants there were bees and others, and they came in handy. Plants therefore created flowers to make them welcome, in order to make them useful. That, by our standards could only have been achieved with a brain behind it. Whatever, there is apparently a hyper-sensitivity present in plants beyond our knowledge, which resulted in such creativity. A flower functions as a magnetic pole for bees, because of their attraction to it, with the colony as its anti-pole. The bee lives between the two stations, which results in a to-and-fro action. The plant made the bee feel at home in its floral world, leaving its own fauna world a beeline away.

The creations that were realised due to the assimilation process were the Very Important Plants in the living world. A plant creating the image of a bumblebee in its flower is perplexing to say the least! No wonder there is plenty to write about to explain plants. They are breakaways. They involve outer space (via sunlight) simply to be.

Light to them has an entirely different meaning of vital importance, totally alien to animals. In plants' case, life and light meet and that is interesting *in optima forma*. It makes Earth come second in this – even as earth (however rich the soil may be). A flower in itself is life blossoming. A flower is awareness put into colour, shape and fragrance. In other words, 'saying it with flowers' reached bees, among others, and 'saying it with leaves' reached tree height!

Plants represent an entire kingdom which is realised photosynthetically. In one sweep it went from light as a force from outer space to light as in our brainwaves or, to put a slightly finer point on the latter part, it went from eyes to see to 'I see', which is a matter from 'greenprint' (leaves) to 'blueprint' (brainwaves). The latter is the more enhanced because leaves are usually paper-thin. In short, as plants have it, life is light made to a very great extent.

Life often works like a magnet and it is expressed in a female moth attracting the opposite sex from a village away, so to speak, or much more powerfully still in a flower attracting members of the world of fauna as a foreign workforce for their 'plant of works'. This powerful pull is also outstandingly expressed in a dog finding its way back home over hundreds of miles. Life shows itself to be capable of bridging huge gaps.

Plants are crucial to the viability of all life on Earth, including Homo sapiens. It could be said in one sweep that life on Earth has been realised first and foremost by 'saying it with plants', while they themselves 'say it with flowers' as the optimum green creation.

Plants are a green, silent explosion or a green big bang on Earth – without a sound! They are a green might through light. Plants were created out of pure necessity to make room for another living world, that of fauna, apart from the fact that they turned out to be a grandiose realm with enormous variety and, above all, they made their own creation – that of flowers.

Light, in other words, is interwoven in the fabric of the universe and light is 'common ground' in the universe. It is through this responding or 'feeling its way' that life has actually been realised on this planet. It is due to this light factor that plants are familiar with outer space.

It may well be put across here also that the bee orchid, as a perplexing example, is evidence that the entire living world has been realised with such a super force behind it.

It is through plants that Earth is linked up with the rest of the universe in a very direct sense. Plant food is light-based in principle, and since we are all essentially chlorophyll-based, directly or indirectly, the living world is therefore made by the sun and earth, with the emphasis on the sun. In other words, the living world as one 'plant of works' consists of two major sources. Since all life is chlorophyll-related, even our brains are light-based in origin – as well as any flower. Plants, therefore, are the 'green middleman' between the abstract world and the tangible one – and enabled us to evolve and survive.

Light, as Sun's might, not only creates our day, in stark contrast to night, which in itself spells out extremes, but there was also apparently a completely different angle to it, if anything, even more important still, namely the abstract part of it being vital to chlorophyll. This creation gave rise eventually to plants up to the size of trees, besides providing a life-show of colourful flowers all over the world. It turned out to be a beautiful world all of its own. It is, in many cases, a very dramatic achievement, because it often involves a very lengthy stalk as the only link between the two, namely the plant and the flowers. A stalk in other words is a fingerprint. It concerns sensitivity stretched to the limit. It all boils down to the fact that plants put light itself in the limelight.

Timber is as much chlorophyll-based as beautiful flowers and the beams from the former support enormous stress, which is house-worthy.

A flower is a plant's top-floor department, which is usually reached by the stalk; the latter functions as a lift shaft. Plants and light in this instance are an inseparable twosome. They concern a heavenly joint effort. Light is furthermore responsible in most cases for flowers' 'opening hours'. Such is the influence of light on the heart of all living matter.

The creation of plants was the first evidence of the existence of a profound awareness in the living world. Bees, among others, are wing power controlled by flower power. Flowers function in such cases as a colourful magnet.

Flowers may be beautiful, but they are no luxury by any means and they merely represent business as usual. There is therefore no true flower show; that is only in our imaginations. Plants not only say it with flower power, which is already extraordinary, but even excel that concept once again in such a mirror image (including its fragrance) as that of the bee orchid. There is a communicating with bees and others as in so many cases by flowers. There is here an obvious parallel with 'brainy' fauna.

Plants come into being in a most peculiar way, by using sunlight in their existence. Yet it is because of them that there is so much life in a flora and fauna fashion on this planet. It means that plants are in touch with outer space. A tree as a skeleton in winter may look like dormancy gone too far.

In the strictest sense, touching a leaf is touching the extraterrestrial. It is this double approach to existing, as in the case of plants, which eventually resulted in the blossoming of their business, expressed in flowers.

There would be no root system, no flower creation, no tree reaching for the sky, no mimicry, without light being there in the first place. Another bit of interesting data concerns plants' flower power, which results in a show of strength, drawing in the crowd, often in the shape of bees. A flower opens and it takes us aback, awe-inspired, witnessing the beauty of it all.

However, it is comprehensible to us only up to a certain degree, because even we ourselves are merely part of it all and not kings either in the kingdom of the living world.

Plants incorporated light in the first place, which was necessary to create a kingdom and, secondly, attracted insects from the animal kingdom to assist in extending their realm. In other words, plants are light-generated, which resulted in leaf power and this in itself generated flower power.

Plants actually manage to attract wildlife all the way from the other world, like bees to their flowers, which get fertilised, while the bees get a reward in the way of nectar. We take it all for granted as just the way of life, but it is in fact a most ingenious way to realise fertilisation. It resulted in an enormous increase in the plant domain. It is basically a matter of deep-down awareness and that is apparently also profoundly represented in the plant world. It ultimately led to us, Homo sapiens.

It is immobile plants, rooted to the spot as they are, which set in motion the world of fauna well known for its mobility. It is a matter of attraction *par excellence* in plants' case, namely in their way of absorbing part of sunlight and also of absorbing goodness from the soil. It is this combination of light and earth which was to be the basis of all life on Earth. In other words, any earthling owes its life to the cooperation of two heavenly bodies, namely Earth and Sun, and plants even prove this to be so in quite a direct sense. All in all it boils down to the fact that there may be life on Earth but the other super-power, the sun, has had a very big hand in it. It just goes to show plants have an appetite which Earth alone cannot fulfil.

Plants come as seeds or bulbs indirectly, in the pre-plant stage, respectively the generative or vegetative way and, either way, can hold their own much longer than hibernating animals. It is, again, a display of further development than in the world of fauna. This registering *par excellence* is expressed as a 'message understood' in the case of plants.

Wild flowers couldn't be more alien looking in comparison to the plant. A striking example of this is the humble dandelion. If the plant in this case has a *dent-de-lion* appearance in its leaves, its yellow biomass on the stalk's top is a sunflower *in optima forma*. It is such a green power, which expresses plants as being the middleman between Sun and Earth. It all hinges, so to speak, on paper-thin leaves usually, which, though almost see-through, were to be the basis for our existence and, indeed, all life on Earth. A leaf is also the greatest complexity ever developed in creation. It is because of such a hyper-sensitivity that it went from plant to flower and from flower to bee.

Plants' approach to life is 'mission impossible' to us – fauna at large. It is this kind of remoteness between two worlds which is of universal magnitude and expresses life on Earth. In a way, there is more to light than meets the eye and plants prove this in a tangible manner from ground-hugging grass to the dazzling height of a tree. Flowers blossom in a huge variety of shapes and colours. It went to the extent that they created a new world, that of wild flowers. It was to be wildlife's third world and it broke away from the green world, but for the slenderest of stalks. The stress supported by stalks is often unbelievable, being able to uphold the colourful head of the actual flower, which in many cases is quite substantial in dimensions. Fatigue is not even reached while swaying in the wind, so beautifully expressed in a long-stalked dandelion, called *Paardebloem* (horse-flower) in Dutch for some reason.

Plants have to stand up to weather and wind all day and night, besides being most intimately involved in part of sunlight. Plants outdoors actually involve the universe through its light. This is one reason why blossoming of plants materialised in flowers. The latter came into being with bees and others in mind. Such flowers are in fact tuning in to members' requirements from the other world, which led to fertilisation of the flowers. Flowers are plants' creations and their blossoming is evidence of their intense awareness. It is the more sensational because the flower is on the whole a very complex structure. The plant seems to have been realised in a planned fashion.

It is a sobering thought that we too are closely related to abstract and, since attraction also plays an important role in our lives, it merely emphasises the abstract quality. At any rate it leads from 'eyes see' to 'I see'. It may be a bit of tongue twisting, but it is true all the same. There again there is that way of putting it by the famous philosopher Multatuli: 'Nothing is entirely true and even that's not true.' We cannot entirely grasp the word 'absolute'. Plants are familiar with the beyond, being light's home, and have a very close relationship with this abstract. Indeed, it is a marriage *par excellence*. It may not be very scientific, but your guess is as good as mine when it comes to getting to the bottom of it all – and plants beat us even here as well by getting to the bottom of it all in root fashion! Plants are absolutely baffling in that they ever came into being. It is a sobering thought that even the most beautiful flowers owe their existence to the humble, paper-thin leaf, which is a solar panel *par excellence*. Leaves and petals are both light-centred, but their functions are worlds apart.

'Pastures green' and 'How green is my valley' are sayings with a green overtone, being light-related, and that is interesting, since green biomass originates not from this world. It means that plants, albeit being rooted on this planet, are nevertheless much more earthlings in a literal sense. In short, the living world is a leaf-life or a light-love story. Plants are legless movers and it comes across in a much more ingenious way than in the case of snakes.

Flora achieved, however, not only spread in a horizontal way, famously known as ground-hugging grass in fields, but also, and just as impressively, the development of trees in forests – and to make the triangle complete, in-depth spread came into being in the enigmatic appearance of the complex flower. Plants represent profound awareness. They materialised accordingly and green biomass comes in a huge variety. Just imagine all the green plants – a flower, let alone a beautiful one, would have caused immense excitement. Now we take it all for granted! In other words, a flower is in fact a sublime example of glorified awareness, often beautifully expressed. Flower power turned plant life into a gigantic realm.

It was to be plants' greatest development, created due to

awareness. A wonderful bee orchid flower, as a perfect copy of a female bumblebee, is evidence in the extreme of what lengths and depths plants went to in order to express themselves – and it works because the male bumblebee 'mates' with it, since the flower has a female bee fragrance. The action of the insect means pollen is spread from one bee orchid to the next one, realising fertilisation. It is this joint effort of the sun's light and the earth's soil which resulted in the creation of huge biomass in the form of flora and fauna on this planet. However, it is especially light which is responsible for life on Earth, because it realises chlorophyll by means of photosynthesis. It is most extraordinary that life at the beginning of its existence used sunlight to make all-important chlorophyll, which was necessary to result in two living realms, that of flora and fauna, and that assimilation process is still going on as strong as ever.

Animals are well known for their leg power on the whole, while plants are anchored to the spot. The difference couldn't be more pronounced. Furthermore, animals generally have bodies whereas plants are see-through. Yet it all happened on one planet! Plants came into being due to awareness and within their realm are the most exciting examples of this abstract, amplified in their creation of flowers. It proves that awareness has been part of the life form from the start and it stands out in issues like mimicry and especially in the development, indirectly, of Homo sapiens. The potentials within were immeasurable, concerning life from the beginning, and the development resulted in two completely different worlds. Plants in this are a 'bridge' between light and life. Flowering plants are indeed a BBC themselves, a Blossoming Broadcasting Centre, firstly because many are capable of 'calling up' members of the other world and secondly as seed-dispensing units. There is in the plant world an utter perplexity, that of bee and orchid put together in a bee orchid. It is a sensational issue by any standard. It is to us completely incomprehensible that it has ever been realised.

Sensitivity is expressed as bonds in plants' case, first and foremost between life and light. There is also a bond between plants and soil, and then there is the more important bond between plants and animals – and to be more precise in a two-fold way, namely plants as a feeding source and secondly plants' flowers attracting animal species, like insects, for their fertilisation process. Plants are fuel for animals, whether directly or indirectly. Plants are in their way proof that life was something very special and not a matter of just growth. It expresses a profound awareness all throughout life, in which mimicry is but an example in flora and fauna. It is this abstract that made life develop in its abundance in a three-dimensional way.

It is well worthwhile to mention here one striking example, namely that of the holly bush, which has no spiky leaves from halfway up, being out of reach of animals' interference. It is in itself an awe-inspiring feature. However, above all in this case is of course the flower, which is a most colourful display of awareness. The potential of realisation was there from the start, which made life a super-might from the moment go. The big question is, of course: what led up to it? Plants' intense intimacy with part of sunlight proves beyond a shadow of a doubt that life is much more that just an earthly power. In other words, a universal might outside Earth has its most important share in realising the living world on this planet. It proves that life is more than just earthly and it involves the 'beyond'. Plants are green due to contact made with outer space. All life depends on it.

Furthermore, all beings great and small, flora and fauna, are realised by awareness. It has been simply part of life from the beginning. Plants absorb light, which is 'channelled' to Earth over millions of miles. They channel via their roots goodness from underground to green growth above, which could be quite substantial in the case of trees. Furthermore, there is in most cases a stalk, which represents a first-class channelling and there is the famous beeline, which is in fact a 'channelling' both ways, and last but not least plants seeds only too often leave home via an airliner, so to speak, in the shape of a bird, which in itself is a kind of channelling.

Plants have their ways of expressing themselves. It is most extraordinary to say the least that relatively small and paper-thin leaves, or just evergreen needles, have been able to create trees the size of huge giants, and that there are ones that are thousands of years old. Our own food for existence is light-based (even beef originates from cattle grazing). Flowers have an existence in which light from outer space has a big share. Photosynthesis also makes a plant blossom – in its flower. Light is apparently much more than a bright sight. Plants in this case are our middlemen *par excellence*. Plants are neither proper earthlings nor entirely green beings from outer space, but rather halfway creations. It is even underlined as such because rooting has a profound anchoring effect. We may be the ultimate kingdom, but it is enigmatic plants that enabled us to develop.

A cutting put in the ground to create new growth in a plant, shrub or tree is just perplexing. It is completely unheard of in the animal kingdom and it expresses plant life as an out of this world power. The procedure couldn't be simpler – but what an effect! It is in itself clear evidence that plant life concerns a power in which it is in close contact with the supernatural known as light. Light and plants are a joint effort, which is most interesting and unknown in the fauna world. It is evidence that plants are semi-extraterrestrial and a cutting proves this, despite the fact that it is a terrestrial contact of the first order. It is nevertheless a display in which universal might has its expression. Another typical example is plants blossoming in flowers. It makes it all well worthwhile to have a closer look at plants. One outstanding display in length is definitely the enormous size of quite a few trees in this world, the more outstanding since it is achieved by the force of simple paper-thin leaves or needle-shaped ones and then of course depth is expressed in the most extraordinary and often very complex display of flowers, again totally unknown in the animal world.

It makes plants most exciting to say the least, and that would be the way of expressing it if sapient extraterrestrial aliens landed here. Plants' most intimate contact with light was bound to show in most extraordinary ways and we see the evidence of it all around. A beautiful flower, a cutting or indeed a bee orchid are all phenomenal enigmas in the living world, and all because of that extreme intimate contact with light. It is a pity we can't penetrate the beyond, but illustrations shown in plant life are quite exciting all the same. It is all taken for granted, while in fact it is well worth looking deeper into it. Light is an abstract worth much more than just daylight and it had a huge share in creating life on

Earth. Both powers of life and light have their base in the beyond and, since plants are realised mainly through sunlight, so much happened in their world of a phenomenal quality – and flowers are just one example. Light's very intimate involvement in existence, realising flora and fauna, means that all life on Earth has extraterrestrial tendencies, the more so since its origin is also out of this world. This is why plants' expressions, especially in flowers, show themselves to be phenomenal.

It is also adaptation which in plants' case had to be so much more a big show, simply because of their restrictions due to immobility, which meant they had to draw potential towards them purely to exist. It resulted among other things in the flower, which moreover in many cases shows an uncanny way of attracting.

It is a sobering thought that all this performing, like blossoming, stinging and taking root among many other instances, is realised by one 'simple' creation, namely paper-thin leaves. Flowers and definitely trees are each in their own right almost a world all by themselves, yet they still belong to the realm of plants and it shows here again that flora went out of its way to express itself. There are no such striking values in the animal world and an elephant is dwarfed by an absolutely giant tree, while there is nothing in fauna as great as a beautiful flower – a world in a world. The extraterrestrial has made its impact on wildlife in general, but it is especially plants which are much more than just influenced by it and managed to have their own creation, as a creation in a creation, namely the world of flowers.

Not only do plants originate from an out of this world environment, but they are in very intimate contact with it through sunlight and, unbelievable as it may sound, it is this abstract, once processed, which realises them. No wonder phenomena happen in their world, and an outstanding example is of course the beautiful flower. In other words a beautiful flower in the world is awareness magnificently and perplexingly expressed.

Flowers' attraction is expressed *in optima forma*, capable consequently of drawing creatures from the other world towards them, usually in the shape of flying insects. It might as well be mentioned twice here, given its immense importance and that not only are such creatures drawn often from a long way away, but, what's more, they represent life expressed in an absolutely different way, with the emphasis on absolutely. It proves again that there is a super power in plants, which is just as well because they need it urgently, anchored to the spot as they are.

It is all very well mentioning all this but the question remains: what led up to it? Life possesses a great deal of depth from the beginning onwards. It seems most unlikely that the two units which express life on Earth and are different in an absolute sense originate from one life core. The differences between the two, in characters and expressions, are worlds apart, and therefore flora and fauna most likely originate from two totally different life cores.

However, to make it more complicated, fauna couldn't exist independently on this planet and needed flora for life, while flora would have been severely restricted without animals' assistance. Mimicry is nature's fooling and that goes for animals as well as the plant world (another striking example of the flora way is the dead-nettle, a lookalike of a stinging nettle, without the stings). This extraordinary concept of 'seeing double' is there for a very good reason and many a species' survival depends on it. This most outstanding image proves here again there is a realising present in fauna as well as flora. It is just baffling, to say the least.

The attractive potentiality expressed by plants in their flowers is much stronger – and here again reaches a quality of the beyond. In plants' case it is a matter of drawing members from a totally different world towards them and pleasing them with aromas they prefer. Furthermore, flowers' colours and shapes have great interest and insects are usually rewarded with nectar once they enter the colourful premises. It all sounds so straightforward, but in fact it expresses altogether an awareness on plants' part in the extreme, since it enables them to create most unbelievably – if it wasn't so colourfully true and provided with a heavenly smell often – something so powerful that it stands out as a world all of its own. Both plant and complex flowers are evidence of all-might.

It is in both cases a matter of expressing awareness, apparent in a super-quality, which lies in the beyond, and we cannot get closer to it. It turns life into an alien invasion and, plant life being twice in touch with outer space, namely in the case of all life's origin and via the assimilation process, stresses the point in particular. That an extremely simple life form not unlike that of a microbe was capable of processing part of sunlight and realising green biomass goes again beyond any description.

Plants are a wonderful world! There was an urge from the very beginning to have a good go at life. The potentials from the moment 'go' were more than just enormous in quantity and quality. Plants with all their extensions, such as their runners, creepers and climbers, express a matter of green life with its additions. They also have thorns and stings as extensions, all unheard of in the animal kingdom.

What actually is a plant? A plant is green biomass that is created by being in touch with sunlight, and processed accordingly. It developed into a huge kingdom and secondary contact is made with the earth's soil via a root system, to reach for goodness within. Realisation masters both kingdoms. The living world, split in two, is one huge organisation and, despite that, no mistakes, no catastrophes, have occurred. This is the more interesting, because they interfere in each other's worlds. Herds graze for one, and plants attract members of fauna. It is the more interesting still because they live in alien fashion. It is most amazing they ever met on one planet. Plants which start off with just about nothing to go by, namely part of abstract sunlight, manage to grow into sky-scraping trees, in some cases with absolutely enormous trunks. It is in its own way a quality which lies in the beyond and the same goes for a beautiful flower. The two creations concern expressions, that of length in trees and that of depth in the case of beautiful flowers which we, the highest developed beings, admire, feeling awe-inspired. What a sublime might is life! In the case of such a giant tree and likewise a beautiful flower, abstract sunlight has been the start of it all.

The quality of abstract had a great share in establishing a living world as expressed in awareness, attraction and sense. Plants merely emphasise that fact in their creation, being dependent on light. In that case it might well have been the case from arrival onwards – if not before contact was made! It means that both living worlds, flora and fauna, owe their existence to the abstract to a very great extent. Furthermore, we all breathe in fresh air (hopefully!) for life, which is yet another abstract. Then when we die, fauna or flora disappear into nothingness eventually. There is also that interlocking apparent as expressed in one very clear example between bee and orchid, and so there is a belonging between the earth and universe. It is perplexing to say the least that a plant had it in itself to copy in the most precise manner the outline and colour of a specific insect – and even has its female fragrance, just to draw the male's attention in a most successful way. It is the more remarkable still since there is no 'grey matter' involved in the realisation. It proves in itself the point that abstract awareness has its outstanding expressions in fauna, glorified in us, as well as flora. Apparently there was a huge urge, plant-wise, to see to it that fertilisation would be realised. What an enormous mental strength within a plant! It certainly isn't a coincidence; a plant went out of its way to perform.

Mention should be made here also of the fact that orchid seeds on the whole are very small indeed – and that concerns the beginning of the above mentioned story with a downright realistic basis to it. It is in itself, as far as the plant is concerned, a matter of moving places, figuratively speaking, while in reality it is anchored due to a rooting system. The plant is also an example of sensing in depth far beyond any expression in fauna that way.

In short, the bee orchid is an out of this world creation in this world. Without the existence of a bee orchid, plants would still have been very interesting in their behaviour, but this puts an even finer point on it. A bee orchid is an outstanding example of what life's all about. A bee orchid is merely a name of ours for a perplexity in the living world, which emphasises the sheer depth penetrated by creation, known as life on earth. It shows there is more to flowering – to plants – and that reaches a quality where words fail. Beautiful flowers in the wild are already very awe-inspiring, but a bee orchid goes one big step further still. It is indeed incredible that plants which have no eyesight and are rooted to the spot have ever been able to make a very clear copy of a member of the other world. It puzzles us deep down as to how it ever was realised. It is a matter of superlative awareness, which is beyond us. It is the more amazing that a bee orchid was ever created, taking into account the fact that plant life's start was very primitive and still is a matter of processing part of sunlight for existence. It puts plants in a much higher order then merely representing a nice countryside and creating interesting gardens. A bee orchid is a sublime example of the mental strength plants apparently possess. Even if plants, just germinated, were provided with sight, it would still be perplexing if one of them were able to copy something in their own realm – let alone creating in a three-dimensional way a member of the other kingdom.

This is all about the inside story of plants and is quite interesting. A bee orchid shows the enormous inner power of plants, but then so are flowers in general a realisation of a creation (that of plants) and in the case of that orchid are expressed as a creation within a creation, unknown in the animal world. Flowers in general, but especially the bee orchid, came into existence most likely because plants' inner strength was released due to a most intimate involvement with part of sunlight, which is an outer space quality. In the case of this orchid flower it is a matter of selecting one specific specimen of the other kingdom to copy, for the sake of attracting it, which in this case is a bumblebee, to realise fertilization of the flower. It is indeed a very ambitious way of doing so – yet it works! It means that there is apparently a power of sight beyond that of eyes, which plants establish. It makes flora at least as great as fauna, strength-wise, having supernatural qualities which we can only admire most profoundly. In a bee orchid plants prove that they have it 'in' them, in an impressive manner. It expresses plant-might in the superlative. This flower expresses wildlife's greatness with its boundaries in the beyond. It would already have been most amazing if plants had in one way or another some kind of sight to be able to copy a member of the other kingdom, but without any sign of vision it is so much the more perplexing, to say the least.

How do plants realise there is another living world? With their creation of complex flowers they certainly prove themselves to be aware of another world. Another big question is, why did plants go to such lengths and depths to multiply?

In short, a bee orchid is an enigmatic power in plant life. It is evidence in an outstanding way that there is more to it. A bee orchid is proof of supernatural existence. It came about by a telepathic force, but then so did the thorn and sting and of course flowers in general. It is most amazing, realising that it all started from minuscule units making contact with Earth and reaching qualities in length and depth of huge magnitudes. It proves there is a mind over matter quality in it all.

To express it all in more detail a bee orchid is a creation in the sixth sense: first of all there was the arrival of life on Earth; secondly, creation of plants; thirdly, multiplying by means of spores as in the case of ferns; fourthly, creation of flowers, the wind-pollinated ones; fifthly, the cross-pollinated ones with the help from outside, namely members of the other kingdom; and last but not least, sixth, the extraordinary creation of a flower which is a copy of a member of the other kingdom. That is action stations in a big way – and that in the 'common' plant world!

It is the more interesting realising the world of flora could have been quite substantial without flowers. It is 'simply' because they anchored to the spot, and they had to 'play clever' to extend their realm enormously. Nevertheless, to involve members of the other world in their way of living, quite apart from the fact they realise there is such a kingdom, shows they are of a higher order. Bee and orchid, as species, are worlds apart, yet amazingly enough there is a bee orchid in the plant world of which there is a great deal mentioned in these pages. It had to be because obviously it is a most amazing specimen. It is one image that makes plants very special as a living world. It sounds so far-fetched, yet it is the real thing.

There is apparently a super-might in plant life, which is per-plexing and surfaces already in beautiful flowers outside, which are made that way to attract fauna. Apparently there is more to life than meets the eye, and plants make doubly sure. They had to, because they can't move and therefore had to come up with intriguing ways to spread their species over the world. It means that there is in the plant kingdom too an inner strength that is just awe-inspiring, without us being able to go into it any further, alas.

Such a force within made the living world, on the whole, what it is worth. It is so much the more amazing, since it wasn't at all necessary to go to such lengths and depths to create a bee orchid. They are an extraterrestrial creation on terrestrial ground. Plants are in touch with the other kingdom via a form of telecommuni-cation, which sounds impossible to us, and a bee orchid just puts a finer point on it. Plants also went to extremes of expressing themselves, as shown in sky-high trees in some parts of the world, in beautiful flowers, which are exciting creations, and by copying an insect, including its fragrance in one of their flowers. They show in the bee orchid the enormous flexibility possible within their kingdom.

Deciduous trees mimic death in wintertime, looking like skeletons, just to hold their own during severe conditions. One of the most outstanding impressions is the creation of colourful flowers and their complexity, all for a very good reason. Plants are a kingdom which has nothing in common with the other living world. Yet due to the fact that plants, lacking sight, are equipped with awareness in the extreme, they are capable of realising the other world and making profound use of it.

It is those kinds of potentials that were part of life when it made contact with this planet. It could be said that this book is quite repetitive, but then that is no problem, really, since it is such an interesting subject, because plants express life in astonishing ways. It is worthwhile writing about it all, considering that plants had such a humble beginning to say the least and are deeply involved in the abstract. It is this kind of life's start that flora and fauna depend on. Evolution followed suit.

There are no less than five different kinds of abstracts involved in plants creation: firstly, it all started from outer space before making contact with this planet; secondly, the famous assimilation process in which abstract sunlight is of vital importance; thirdly, the abstract involvement of intense sensitivity; as point four, the abstract of vital importance, that of awareness; and last but not least, attraction as a major part of the creation of flowers.

It is amazing that fruit like an apple increased hugely in size over the aeons due to our interference, compared to its original dwarf size like that of the crab apple. However, its seeds, known as pips, with all sorts of data, remained about the same size. Apparently that was the correct size – and still is – for transport by animals, to be deposited far away from the original tree.

It is awareness and sensitivity, both in a very profound way, which made the living world what it is worth, picking up data as they went along. Flowers, especially the beautiful ones in the wilds, are an excellent example as to what could be achieved. It is the more interesting because the flower, in all its varieties, wasn't at all necessary to be created. The same can be said of the outstanding developments in the fauna world. It all could have been left to primitive life, but precision work took over and consequently we arrived.

There was apparently more behind it all, and the way plants have expressed themselves borders on the unbelievable. They had plainly an inner strength to make a huge show on Earth. Plants expressed themselves beyond expectations. It is extraordinary it ever came to realisation. It shows plants to be talented, where wavelengths within show the awe-inspiring. It obviously should impress us tremendously!

Plants, as biomass, were the first on the scene, which is just as well, because that lead to a viability for fauna. It turned it all into a reasonable state of affairs and this situation was already a fait accompli from the very beginning. Flowers are evidence that plants went a much longer way, especially in depth, than was strictly speaking necessary. In short, flowers are an extra – and the big question is: why? There is a force within plants that reacts in a mind over matter way. The two living worlds have that in common. There is a profound responding in fauna and flora. Once again: What is a plant? A plant is green biomass which is realised in a most enigmatic way, namely due to processing part of sunlight, resulting in chlorophyll, which is of vital importance to plants. Furthermore, they depend on a root system, which is of secondary importance for their existence.

It is the more extraordinary that the very complicated concept of a flower came into existence, because ferns and mosses among others are quite satisfied without any such display, and that could have been the general idea all over the plant kingdom. Something was triggered off within plants to turn into blossoming, however far-fetched, as a *plan de campagne*. It turned out to be most successful and is the reason why their kingdom increased in size enormously. It proves quite clearly that there is much more to life than simply plant or animal!

Plants show what they are really worth in their flowers especially, and we can only admire. It shows plants are provided with an inner force which words fail to express. It even came to copying in their flowers, of which the bee orchid is a splendid example. In short: a flower is a display by plants in a most mysterious way, which words fail to express completely. In other words, a flower was abstract once, millions of years ago, and turned tangible due to profound sensitivity and awareness. It is often beautiful three-dimensionally, namely in colour, outline and depth, besides being provided often with a most pleasant fragrance, alluring members from the other world.

It is again all a matter of reasoning or, in other words, putting two and two together. Profound sensitivity concerning a quality which just baffles us surfaces in examples such as a dog finding its way home over hundreds of miles or swallows leaving their winter quarters in Africa and flying back home all the way to the Continent or Great Britain for the sake of nesting (which is just puzzling in its own right – why not just stay put?), or, indeed, a burning issue like the leaves of a stinging nettle provided with very delicate, glassy, brittle stings, which break upon touch and release a poisonous fluid just for the sake of defending themselves.

Plants are the link with outer space in view of their very intimate involvement with part of sunlight, being a source of force from the other heavenly body. The assimilation process is still a very big undertaking to realise plants at present, let alone the arrival of life on Earth – and it proves again that deep down awareness had a big share in it all. It is part of life and always has been. Life, therefore, is not just tangible matter but also of an abstract nature. Consciousness is shown in a most stunning manner in plants' fashion, and a bee orchid tops it all. Plants are created by means of processing sunlight, which is a very complicated action in the living world at present, let alone soon after the arrival of life on Earth! It proves the enormous power which was involved to create two living worlds.

Plants came into being in a most unlikely manner, involving abstract matter in outer space for their existence. The sensitive plant shows awareness in a perplexing way by folding up its tiny elongated leaves on one stem within a very short space of time when touched. Plants' realm was already quite substantial at first, consisting of ferns, including tree ferns and mosses among many others. That was their pre-flowering period and their world increased dramatically once it came to blossoming, which is evidence of profound awareness again, combined with super-sensitivity.

Flowers in general developed entirely in a world all of their own, often provided with long stalks for a good reason. The variety in flowers is enormous, in colour, size and shape, and they are often provided with a heavenly smell, all with multiplication in mind.

What stands out above all is the force involved, beyond description, that made it all come true. Plants' pre-flowering period was already mysterious in view of the creation of chlorophyll for life. It is, alas, all beyond us and we merely see the evidence of it, and are awe-inspired. It is worth mentioning here, once again, the utter sensation it would have caused if someone, more or less in our time, had seen the first plant with a flower deep in the wilds. We could surely then have concluded there is a mind over matter in the plant kingdom as well, most astonishingly. Plants are, therefore, capable of realising they exist and making the best of it. In other words, plants too 'tick over' so to speak and there is a registering apparent in both worlds.

Plants meant the possibility of life to animals, while the plant world itself increased dramatically in size due to animal activity, especially that of insects. It has its own climax in the bee meeting the flower. Plants sense the animal world, a sense which is strikingly demonstrated in their colourful flowers. It could also be said that plants are the green vein within all of us. Their variety is enormous and even a skeletal tree in wintertime, a contradiction in terms *in optima forma*, is still leaf-originated.

Plants also created insect traps in their leaves, for the sake of feeding on them. The structures are very cleverly made and it is, again, clear evidence of awareness in plants. There are quite a few different kinds. In short: plants are aware; they realise. It also made them what they are and evolution merely followed suit. Their green world experiences a huge handicap in immobility which is underlined by their being rooted to the spot, and this is the reason why there are many counteracting displays like beautiful flowers and thorns among many others.

In a sense, plants renewed contact with the sun, while the earth broke away from this heavenly body. In a sense, plants read life in a foreign language that we cannot translate entirely. All plants are sensitive and the actual sensitive plant in this is but an instance. They are all leaf-made, which sparked off flowers eventually and both creations are shrouded in mystery. In short, plants concern a vision beyond sight.

This green world also concerns 'The Involvers', because they involve the ground they stand on, they involve the other kingdom in their livelihood, but above all, they involve sunlight – or at least part of it – to exist. There was to be life on Earth in a two-fold way, due to this process of photosynthesis carried out by plants. It even came to flowering that way, which is also beyond expectations.

In a sense, animals are as natural as plants are supernatural. Plants just improved over the ages and became a biomass of a huge variety, but basically they are all the same, depending on chlorophyll to exist. They have done so from close to the very beginning of life on Earth, yet their variety turned out to be nevertheless enormous, which is the more amazing because they are all anchored to the spot in principle. Not only that, but they managed to create a new world within them, that of flowers, which themselves consist of a huge variety, often emphasised as a realm all of their own by means of a stalk, which serves as a flagpole so they can be seen from far away. It is all a matter of reasoning by plants. To plants, light, or at least a very specific part of it, has a much more profound meaning. It has a major share in their creation. Without it there simply wouldn't be any plants – or, consequently, animals.

Light, therefore, has a two-fold meaning: firstly, objectively, and secondly, subjectively. In short: light means life. It makes the sun as important as the earth for life. It certainly has a major share in all life, like attraction and awareness. It also comes down to the fact that plant life is light-centred and we are all plant-bound. It has resulted in the fact that plants are the green manna from heaven. Plants are a mind over matter issue. There is a huge difference in that make in contrast to the other kingdom. An elephant may rest in the shade of a tree, but the two are nevertheless millions of miles apart as forms of life, metaphorically speaking. Plants' kingdom came into being without a head between them, no stomach, no pair of lungs, no legs, no heart and no eyes to see, yet they created a world hugely admired by us!

One of the most outstanding features of life on Earth is undoubtedly the concept of mimicry, which has its perplexities in both worlds. It has in common with chlorophyll the fact that there is a sensing apparent in the superlative. Mimicry is a matter of 'seeing double' without necessarily a pair of eyes being involved. It expresses to us a 'beyond expectations' effect. Furthermore, it is out of reach for evolution. A mimic is no gimmick, but a feature for life's sake.

Leaves are where flatness is not superficial. Due to leaves, so much came to blossoming, literally and figuratively. They are a green line followed up with perplexing results.

Flowers are often a 'beautiful face' – without an opposite sex in their species admiring it. There again, flowers are often a sweet shop with special opening hours – for customers from the other world. It results in a linking up of both realms. To put it in another way, flowers are plants' 'brainchild'. Abstract has such a huge share in it all and the flower, millions of years ago, was realised in a very primitive way due to awareness by plants. Even a baby owes its being on attraction in the first place of two individuals of opposite sexes, and, consequently, we get the impression that life itself, when it all started, owed its creation to attraction to a very great extent.

Plants are attracted to light, which originates in outer space. Later in life plants in their way attract life all the way from the other world and, therefore, represent a far-fetched concept. Plants, in other words, live a life which borders on the unbelievable and we can read it only up to a point. Flowers make sense and they express awareness. A whole new range of green developments came into being due to colourful flowers. They are more unique as a creation than we are. We merely followed suit! It is all due to the fact that light, a specific part of it, is greatly responsible for life on Earth. Plants read the abstract light much more profoundly than we ever will. Due to that know-how, we are here too. In other words, life is a matter of highlighting light. Plants have shown themselves to possess huge stamina to overcome their handicap of being immobile, as is especially starkly displayed in their colourful flowers. It is this consciousness that rules the living world.

The enormous variety in fruits is another clear example of plants' (trees and shrubs mainly) profound awareness, and it results in the seeds being deposited a long way away on purpose. Plants on the whole are three-dimensional in their creation, namely as seed, chlorophyll and their root system, but a fourth dimension could be mentioned: the fact that most of them flower, which is an extraordinary creation and meant a huge extension to their realm and, again, it is an expression of profound awareness. It shows plants' intense involvement to assure life on Earth. Their assimilation process shows that they are nevertheless only partially earthly. It is amazing to say the least that all this started from so little, which in itself is a matter of extremes. It is this most extraordinary processing of matter that has had a major share in creating plants.

The following pages contain a range of definitions, as a summary of plant life.

## *Plants*

1.  Plants are realised due to abstract awareness and abstract light as a twosome.

2.  Plants, in other words, are an enigma of the first order.

3.  Plants are a profound penetration in light and earth.

4.  Plants more than just border on the unbelievable.

5.  Plants rely on abstract light which, processed, turns into tangible green matter.

6.  Plants are mysterious.

7.  Plants are leaf-based. Their leaves are usually paper-thin and led to the most amazing expressions, like beautiful flowers and plants the size of sky-high trees, so gigantic they have to be seen to be believed.

8.  Plants are the outcome of two heavenly bodies cooperating.

9.  Plants make (Sun) light work!

10. To plants, light means life.

11. Plants, as a species, are the unthinkable come true.

12. Plants, which had a very primitive beginning, were nevertheless created due to the most complicated process of photosynthesis, realising chlorophyll.

13. Plants are a matter of penetration, of sunlight and Earth.

14. Plants' awareness didn't stop at the multitude in variety from grass to tree size among others, but went into depth on a huge scale as shown in flowers, metaphorically speaking, especially in the very complicated types such as orchids, showing a quality of awareness which is the equivalent of our brain power.

15. Plants, in other words, show themselves to possess a degree of sensitivity, which can only be compared with the well-

developed creatures of the other kingdom as far as the sophisticated types are concerned. It is the more interesting in plants' case because they had such a very humble beginning – which still exists today!

16. Plants' approach to life is different to that of animals' in the extreme.

17. Plants, in principle, remained the same throughout the ages and, consequently, a seedling is totally dependent on part of light to survive – just like green biomass started all that time ago.

18. Plants show clearly how very important this abstract light is, both subjectively and, above all, objectively, to the extent that all life, split into two worlds, vitally depends on it.

19. Plants, as green biomass, were preceded by a life unit, which somehow realised the vital importance of part of light and due to that fact not just plants but we too are here! In short: we owe our life to light.

20. Plants express a biography well worth reading.

21. Plants made evolution possible.

22. Plants' attraction to light was a matter of awareness in the extreme – and still is.

23. Plants went for the sheer out of reach option to realise life on Earth.

24. Plants are created by chlorophyll and responsible for a huge variety on the theme, which led to the creation of flowers and a totally different world, that of fauna.

25. Plants exist due to the most enigmatic assimilation process, which doesn't make one any the wiser.

26. Plant life's origin, in other words, is quite puzzling, but it held its own over the ages, thank goodness.

27. Plants – the big question is: what spurred life on at the start to tackle a value of such a huge dimension?

28. Plants show plenty of forms of awareness and act accordingly, especially in all the different types of orchid flowers.

29. Plants' creation is largely abstract involvement.

30. Plants' own creation is the flower, which is an expression of profound awareness.

31. Plants themselves are a creation within a creation, since very primitive green biomass like that of algae is barely recognisable as plant life, whereas trees are the other extreme, and it is hard to believe they are part of plant life. It is, again, a matter of extremes as far as plants are concerned, being in very intimate touch with part of light and, on the other hand, penetrating thoroughly in terra firma.

32. Plants may be stationary in principle, but they have their ways to overcome this handicap to a certain extent, namely by means of creeping, running and climbing.

33. Plants' creation is beyond description, for the contact with part of sunlight is too intimate.

34. Plants were a necessity to realise two living worlds, however far-fetched the idea.

35. Plants changed the surface of the Earth for ever, despite the fact that many plants are annuals.

36. Plants have yet another way in store to survive, namely by making ends meet as roots (including shapes like bulbs), whereas animals have to satisfy themselves by merely fattening up before hibernating or growing a thicker pelt pre-winter.

37. Plants are emitters (of seeds) and receiving stations (of bees and others) due to awareness.

38. Plants owe their life to light, primarily, and secondly to roots.

39. Plants went wild expressing themselves.

40. Plants could have been replaced by nothing else to realise two living worlds on Earth.

41. A plant is awareness realised.

42. Plants – we owe our lives to them, however indirectly.

43. Plants are a mixture of soil's goodness and solar energy.

44. Plants led the way to bees and eventually us.

45.  Plants depend on absorbing light, bees and others in their flowers, and goodness out of the soil they stand in.

46.  Plants are a joint effort, or a two-fold approach.

47.  Plants penetrate all the way into the other kingdom objectively, by means of attraction, while the ones attracted penetrate the green world subjectively.

48.  Plants made life on Earth go, stationary as they may be themselves. It is the best contradiction in terms of them all.

49.  A plant is proof that earth alone is not enough to create.

*Leaves*

1. Leaves' origin is abstract-based.
2. A leaf is the basis of all life, directly or indirectly.
3. A leaf is evidence that life possessed huge potential.
4. A leaf is green (usually) for a very good reason, however enigmatic the creation.
5. Leaves are the link between Earth and outer space (namely part of sunlight).
6. A leaf is realised photosynthetically, a matter of changing abstract into tangible matters, which is beyond us.
7. A leaf is where the supernatural turns natural.
8. Leaves may be absent in the skeletal period of trees in wintertime, but trunk and branches are realised by them indirectly all the same.
9. A leaf is where two worlds meet – Sun (its light) and Earth (its soil). It represents two extremes at the same time.
10. A leaf is life as a super-power, realised consequently.
11. A leaf is a life-green image, concerning a creation that excels any description, other than expressing it in a very elementary way.
12. A leaf is the outcome of a most extraordinary processing, involving outer space.
13. Leaves are usually superficial in make, but nonetheless have enormous depth, metaphorically speaking, and led to flowers and trees.
14. Leaves are green data-rich chips.
15. Leaves mean life to wildlife; without them there wouldn't have been any.
16. Leaves are wildlife's green potential, directly or indirectly.

17.  A leaf is the 'green might' that we all depend upon.

18.  Leaves are the focal point all life pivots around.

19.  Leaves are responsible for life upon Earth.

20.  Leaves are a 'godsend'.

21.  Leaves are born by photosynthesis, in which wavelengths have a big share. It comes close to wishful thinking – but so does mimicry!

22.  Leaves are green manna from heaven.

23.  Leaves are usually flimsy, yet are the starting point to so much solid matter, like tree trunks and, indirectly, elephants, among other things.

24.  Leaves are where abstract meets matter.

25.  Leaves were an option which seemed to be out of reach.

26.  Leaves are a variety showing green.

27.  Leaves are the Capability Greens.

28.  Leaves owe their life to light.

29.  Leaves link up two worlds – Sun and Earth – and they are at the same time responsible for creating directly or indirectly two living worlds on this planet.

30.  Leaves are as common as they are uncommon, which still puzzles scientists today. Leaves are bodies' opponents.

31.  A leaf made life what it is.

32.  Leaves are the other option, however seemingly unlikely.

33.  Leaves are as 'far-fetched' as mimicry in its field.

34.  Leaves are at least as precious as the body, if not more so.

35.  Leaves are the outcome of a perplexing processing action, and we wouldn't have been here without this involvement.

36.  Leaves' creation makes sense, since two living worlds depend on it.

37.  A leaf is where assimilation is responsible for so much in the process, creating two living worlds consequently, including that of fauna indirectly.

38. A leaf's creation involves outer space, namely part of sunlight, to realise chlorophyll.

39. Leaves' origin is light-based, and all life depends on it, directly or indirectly.

*Flowers*

1.  A flower is the only way to 'bridge a gap', plants being unable to deliver the goods themselves. It is also seeds' starting point.

2.  A flower is often quite colourful, but not to please its mate (the opposite sex).

3.  A flower is a world all by itself and stressed that way through its long stalk.

4.  Flowers show the true cost of living. They are plant's creation, which is remarkable to say the least in the well-developed kinds.

5.  A flower is the third dimension of a life unit. The other two are roots and leaves, unknown in the other kingdom.

6.  Flowers are exciting if we consider the fact that they were even realised at all.

7.  Flowers have been realised due to plants' awareness.

8.  A flower is where three don't make a crowd.

9.  Flowers even attract us, but the dimension is totally different, in comparison to bees and other insects.

10. Flowers are often a beautiful 'face', without an opposite sex admiring it.

11. Flowers are life blossoming, often beautifully.

12. Flowers are 'beyond expectations'.

13. Flowers are plants' 'eyes', 'seeing' bees, blind as they are!

14. Flowers must have once been 'wishful thinking', which materialised for a good reason.

15. Flowers have a sex-life with a detour and either wind or animals are required and, therefore, three does not make a crowd in the plant world.

16.  A flower is mind over (green) matter.

17.  A flower is awareness flourishing.

18.  A flower is a blossoming beyond green growth and a great variety have others in mind to assist in their sex-life.

19.  Flowers opened up a whole new development in the plant world.

20.  A flower is a tell-tale, realising plants' inner strength.

21.  A flower, however complicated, owes its life to a plant.

22.  Flowers are the unexpected realised. Ferns and mosses, among others, were already quite interesting, even up to tree size in ferns.

23.  Flowers are where the green line turns colourful.

24.  Flowers are plants' mastermind.

25.  Flowers are wild – but only in our terms. We are running out of words, apparently!

26.  Flowers, the well-developed kind, attract insects – and us, all the way from the other world, although the reason behind our attraction is quite different.

27.  Flowers are evidence there was more to come.

28.  Flowers are a display, literally and figuratively, and especially the well-developed kinds show depth well worth looking into.

29.  Flowers are an expression of plants' inner strength.

30.  Flowers are often provided with stalks, which are long necks for a good reason.

31.  A flower is the realisation of plant life exploring.

32.  A flower is a creation within a creation (that of plant).

33.  A flower is blossoming in plant fashion.

34.  Flowers are the outcome of an urge with the purpose to increase the plant world hugely in size.

35.  Flowers were the result of insight.

36.  Flowers are more unique as a creation than we are. We merely followed suit, basically.

37. Flowers are an extension of the concept that survival is of the fittest.

38. Flowers demonstrate development further along.

39. Flowers are a model without a prototype – ferns' fronds can hardly be considered a starting point.

40. Flowers and roots are much greater extremes than heads and tails.

41. Flowers are as delicate as they are exposed.

42. Flowers often have their opening hours, as any good sweet shop does.

43. Flowers are often quite a show and meant that way, with plants as the green driving force behind them.

44. Flowers concern a spread to otherwise immobile plants, involving fauna. It is one way for plants to spread.

45. Flowers' predecessors are leaves and in their creation abstract plays a major role.

46. Flowers are often a show of splendour that the true owner, the plant, can never see!

47. Flowers were plants' way of going further afield.

48. Flowers are a show on Earth that the universe can be proud of!

49. Flowers, it was realised, were to be the answer to increase the plant realm dramatically.

50. Flowers are a clear sign that awareness must have been part of life all throughout its existence. It led eventually to us. It also led to flora and fauna. Evolution just followed suit.

51. Flowers are the flourishing of plants' own 'plant'.

52. Flowers are the crown on wildlife's work.

53. Flower power originates from green power (plants), itself created through the assimilation process, which is light-based.

54. Flowers are an expression of consciousness shown by plants in a most exciting way.

55. Flowers are just baffling, having been created by plants.

56. Flowers show the richness of plants' inner strength, which borders on the unbelievable, especially as far as the truly complicated flowers are concerned, like orchids.

57. Flowers show in a blossoming manner the enormous potential within plants.

58. Flowers' origin was a search for more expansion in the world of plants, and this expression of consciousness has been quite successful.

59. Flowers are realised indirectly due to light, which has a big share in creating plants.

60. Flowers, as a creation, are on top of a creation already.

61. Flowers are indirectly light-related, being leaves' source of origin to a very great extent.

62. Flowers' origin is plant life, which is light-based.

63. Flowers are a show in which beauty is a mere extra.

64. Flowers' functioning is perplexing, often attracting members from the other kingdom for a very good reason, showing plants' potential.

65. Flowers are usually exciting, to say the least, as a creation of the plant world, which itself expresses a most extraordinary origin.

66. Flowers' creation turned out to be a gigantic undertaking, especially in the more complicated types.

67. Flowers realised crops like wheat and rice, which are of vital importance to us.

68. Flowers show plants had the inner strength to realise them.

69. Flowers are a show which plants never see for themselves.

70. Flowers are on top of an already supernatural growth, that of plants.

71. Flowers, the well-developed kinds, are architecturally and artistically speaking very great creations, and this is emphasised in their long stalks.

72. Flowers are mostly created with attraction in mind.

73.  Flowers are evidence of superlative sensing.

74.  Flowers are plants' creations, which involves a super-quality in the well-developed kinds.

*Seeds*

1. Seeds express travelling in the plant kingdom, directly or indirectly.
2. Seeds are a show of strength where dimensions are often minute in the extreme (like in begonias and orchids).
3. Seeds in some cases have the ability to germinate in the most unlikely places and to turn into fully-grown plants, including shrubs or even trees (the buddleia is a very clear example of this).
4. A seed is followed by a silent explosion in green.
5. A seed is as tough as a seedling is tender. It weathers well usually and could turn out into a sheer everlasting green plant.
6. A seed shows often what can by done from barely more than scratch.
7. A seed goes by air, one way or another usually, and it is like a spaceship crammed with data.
8. A seed is perfection in its smallest dimensions.
9. Seeds do the moving along, while it is the reverse in the other kingdom.
10. Seeds home in on earth (with any luck!) after their air travel.

## *Roots*

1. Roots are at the other end of the pole, representing plant life.

2. Roots are an underground pipeline system of vital importance.

3. Roots literally penetrate into matter called earth (of Earth).

4. Roots are plants' invisible strength, so much relied on.

5. Roots are plants' cellar, full of goodness.

6. Roots are the 'other side', showing that plants have it both ways.

7. Roots anchor life on Earth, directly or indirectly.

8. Roots show that underground plant-life is also of vital importance.

9. Roots cause plants to be stationary twice over so to speak, since they are already legless, though running-grass for one may try to contradict that statement.

10. Roots are a pumping station and a tree alone needs gallons of water daily in summer just to quench its thirst and, secondly, to deliver goodness out of the ground it stands on.

11. Roots are as alien to animals as chlorophyll.

12. Roots and chlorophyll show what a plant is really worth.

13. Roots are part of a triangle in the more developed types, namely leaves, roots and flowers.

14. Roots are where life on Earth has taken root.

15. Roots were the other option in comparison to chlorophyll, however far-fetched.

16. Roots are the root of the matter, literally, in plants' case.

*Mimicry*

1.   Mimicry concerns the molehills on an already mysterious field.

2.   Mimicry puts the sensitive plant in the shade.

3.   Mimicry is a spitting image expression without a twinning involved.

4.   Mimicry is a vivid imagination or wishful thinking come to life.

5.   Mimicry concerns a tuning in, but the pitch is unknown to us.

6.   Mimicry is out of reach of evolution.

7.   Mimicry is like mysterious chlorophyll – a sensing in the superlative.

8.   Mimicry is seeing double, often without a pair of eyes being involved.

9.   Mimicry is another branch in the big department store, and nobody knows who built it.

10.   Mimicry is like an identikit come to life.

11.   Mimicry is a fooling, seemingly larger than life and more than once startling enough to make us laugh, as the only remaining way to express ourselves about the incredible precision work involved in, for example, a green moth looking like green moss as it rests itself.

12.   Mimicry is a kidding in earnest and all for the sake of survival.

13.   Mimicry is a photocopy in a three-dimensional way.

14.   Mimicry is putting up an appearance twice over, but the two readings are quite different.

15.   Mimicry is a matter of telecommunication.

16.  Mimicry means an enigma twice over, since life itself is already a puzzle.

17.  Mimicry is where the abstract meets the tangible – which is also the case in the creation of chlorophyll.

18.  Mimicry is a spitting image expression and its effectiveness is understood and utilised.

19.  Mimicry is merely another tell-tale that there is more to it.

20.  Mimicry is a matter of awareness beyond the extreme.

21.  Mimicry is as supernatural as a cutting in the ground growing to a full-sized plant, shrub or tree.

22.  Mimicry is just baffling to say the least, especially in the case of a bee orchid resembling a bumblebee, which even provides the fragrance of a female bumblebee and is all the more perplexing bearing in mind that plants have no vision.

23.  Mimicry is a copying between two totally different images which are, in some cases, worlds apart.

24.  Mimicry is a total mystery, but then we can read life only up to a point.

25.  Mimicry is a duplication with a hallmark unknown to us.

26.  Mimicry is where existence shows itself to possess an extra stretch.

27.  Mimicry's sole purpose in life is to increase vitality.

28.  Mimicry is one step higher on the ladder of development.

29.  Mimicry is proof beyond doubt that it is such a super-strength which also creates plants.

30.  Mimicry shows life one stretch further along for the sake of existence, but the expressions are more than even we, Homo sapiens, are able to explain.

31.  Mimicry gives the impression that it was entirely unnecessary to go to such lengths, until life shows clearly the contrary.

32.  Mimicry is supernatural, naturally.

33.  Mimicry turned out to be very successful as a novelty, but then no mistakes have ever been made in nature.

34. Mimicry is an expression of vision beyond the use of eyesight.

35. Mimicry is no coincidence and it is, in fact, a similarity between two living images on purpose.

36. Mimicry is mind over matter in the superlative.

37. Mimicry is a mirror image as an extraordinary creation, all for the better.

38. Mimicry is evidence of life's boundaries stretched out in the supernatural.

39. Mimicry shows the sheer super-strength of life's creativity potential.

40. Mimicry is the unpredictable and consequently the mystery surfaces which is present within all life.

41. Mimicry is proof that nature's boundaries lie in the beyond and the leaf is just another outstanding example in that respect.

42. Mimicry comes to life almost entirely in the abstract way – so do leaves, which we entirely depend on, indirectly.

43. Mimicry is a novelty on a huge scale, which paid off much to the advantage of the living world.

44. Mimicry is an extra in creation, which resulted in a dramatic increase in the variety of life's species.

45. Mimicry is precision work of a different kind in the case of animals and, indeed, plants.

46. Mimicry cannot be explained away by us despite our being Homo sapiens. It is on a higher plane and out of reach.

47. Mimicry expresses the sheer unbelievable strength of the life-might within.

48. Mimicry is a tuning in of a higher order, beyond our knowledge.

49. Mimicry appeared out of the blue. It increased variety in wildlife, fauna and flora, substantially.

50. Mimicry is far-fetched and its creation has us puzzled, yet it is quite successful all the same.

51. Mimicry is one better still than camouflage, as is especially clearly shown in the bee orchid. It is where plants made a wise decision to benefit by it.

## *The Assimilation Process*

1.  The assimilation process is a photosynthetic happening, realising life on Earth, directly or indirectly.
2.  The assimilation process proves that there is more to plants than just Earth can offer.
3.  The assimilation process is one up for plants, leaving animals behind.
4.  The assimilation process is one that all life on Earth depends on.
5.  The assimilation process involves the beyond in living.
6.  The assimilation process is where the abstract meets the tangible.
7.  The assimilation process realises chlorophyll, which no life can do without, directly or indirectly.
8.  The assimilation process shows the huge impressiveness of life.
9.  The assimilation process has an impact on the living world.
10. The assimilation process is in itself part of the creation with its realisation of a dual living world.
11. The assimilation process was the outcome of a sensing in the superlative. So was mimicry.
12. The assimilation process is responsible for the realisation of chlorophyll, which all life depends on directly or indirectly.
13. The assimilation process was (and still is) an action with an enormous impact.
14. The assimilation process had to happen to open the sluice gates, so to speak, for a huge worldwide spread of life.
15. The assimilation process, however hugely important, is nevertheless an enigma in principle.

16. The assimilation process is the most important event that ever happened in the living world and even we owe life to that concept, however indirectly.

17. The assimilation process is where fantasy turns into reality.

18. The assimilation process puts the accent on the greatness of life on Earth.

19. The assimilation process expresses a force within, which has no equal.

20. The assimilation process was a most unlikely state of affairs, yet absolutely necessary to make life happen.

21. The assimilation process gives the impression that if the abstract played such a major role to realise it all, then life itself might have had its share in the abstract.

22. The assimilation process sounds like an issue too far-fetched, but it turned out to be of vital importance overall.

23. The assimilation process is a happening where we run out of words to express the magnitude fully.

24. The assimilation process is an act that goes from light to chlorophyll most extraordinarily.

25. The assimilation process turned imagination into reality in one specific way.

26. The assimilation process is evidence that super-sensitivity was already part of life at the beginning.

27. The assimilation process was the missing link. From then on progress was assured up to the extent of two huge living worlds all over the planet. Somehow this was realised most effectively, and it even led to us, Homo sapiens.

28. The assimilation process created life on Earth from the life unit(s) that arrived here.

29. The assimilation process was there close to the beginning and is still going strong today.

30. The assimilation process is a feature out of this world, with enormous consequences.

31. The assimilation process is where abstract turns tangible.

32. The assimilation process makes one wonder what mighty stimulus lay behind it to go to such lengths to realise two living worlds on Earth.

33. The assimilation process was to have the biggest impact to realise flora and fauna.

34. The assimilation process set life in motion.

35. The assimilation process met a world which was ready for it.

36. The assimilation process gives the impression that everything was all pre-planned. The fact is it came in very handy at the right time.

37. The assimilation process was a matter of so much developing from so little.

38. The assimilation process was the solution to achieve a worldwide spread, but what an effort, which was to include most importantly abstract matter all the way from the sun.

39. The assimilation process was an out of this world connection as a necessity to create two living worlds on Earth. In short, life is abstract-bound to a very great extent.

40. The assimilation process is in itself again evidence that it has all been realised due to awareness and hyper-sensitivity.

41. The assimilation process shows that the sun is primarily of vital importance to life on Earth, for the sake of green biomass and fauna and, secondly, for light in general and warmth.

42. The assimilation process is an involvement in sunlight of all things and the performance leads to chlorophyll, most amazingly. It is an act of surrealism of the highest order, well beyond our knowledge.

43. The assimilation process's origin was the most sensational event, without involving evolution, that happened in the living world, and it still occurs on a daily basis all over the world.

44. The assimilation process was (and still is) the most unimaginable act in the living world, which all life depends on.

45. The assimilation process sounds so straightforward, while in fact it is the most enigmatic and, at the same time, the most important action ever recorded, which realised two living worlds on one planet.

## A Closing Thought

All this has been a very close look at plants – and they deserve it. So much is going on in the green world that it was considered well worthwhile to examine it closely in this book. It is amazing to say the least that profound awareness and super-sensitivity realised green biomass, which eventually led to the world of plants. It turned out to be that this most intimate contact, which involves part of light, led to a kingdom with the most outstanding appearances, as expressed in flowers and trees – and the latter have their true giants among them. As far as flowers are concerned, they went again from a very humble beginning, as in the case of plants themselves, to the most complicated and striking examples, and the variety in beautiful flowers is enormous. Plants absorb the ground they stand on and, at the same time, absorb the goodness in part of sunlight. It demonstrates the most peculiar existence of plants in which abstract plays such a most important role.

9 781844 018475